LONDON MATHEMATICAL SOCIETY LECTURE NOTE SERIES

Managing Editor: Professor N.J. Hitchin, Mathematical Institute,
University of Oxford, 24–29 St Giles, Oxford OX1 3LB, United Kingdom

The titles below are available from booksellers, or, from Cambridge University

London Mathematical Society Lecture Note Series: 325

Lectures on the Ricci Flow

PETER TOPPING
University of Warwick

CAMBRIDGE
UNIVERSITY PRESS

CAMBRIDGE UNIVERSITY PRESS
Cambridge, New York, Melbourne, Madrid, Cape Town, Singapore,
São Paulo, Delhi, Dubai, Tokyo

Cambridge University Press
The Edinburgh Building, Cambridge CB2 8RU, UK

Published in the United States of America by Cambridge University Press, New York

www.cambridge.org
Information on this title: www.cambridge.org/9780521689472

First published 2006

A catalogue record for this publication is available from the British Library

ISBN 978-0-521-68947-2 Paperback

Transferred to digital printing 2009

Contents

Preface

These notes represent an updated version of a course on Hamilton's Ricci flow that I gave at the University of Warwick in the spring of 2004. I have aimed to give an introduction to the main ideas of the subject, a large proportion of which are due to Hamilton over the period since he introduced the Ricci flow in 1982. The main difference between these notes and others which are available at the time of writing is that I follow the quite different route which is natural in the light of work of Perelman from 2002. It is now understood how to 'blow up' general Ricci flows near their singularities, as one is used to doing in other contexts within geometric analysis. This technique is now central to the subject, and we emphasise it throughout, illustrating it in Chapter 10 by giving a modern proof of Hamilton's theorem that a closed three-dimensional manifold which carries a metric of positive Ricci curvature is a spherical space form.

Aside from the selection of material, there is nothing in these notes which should be considered new. There are quite a few points which have been clarified, and we have given some proofs of well-known facts for which we know of no good reference. The proof we give of Hamilton's theorem does not appear elsewhere in print, but should be clear to experts. The reader will also find some mild reformulations, for example of the curvature pinching results in Chapter 9.

The original lectures were delivered to a mixture of graduate students, postdocs, staff, and even some undergraduates. Generally I assumed that the audience had just completed a first course in differential geometry, and an elementary course in PDE, and were just about to embark on a more advanced course in PDE. I tried to make the lectures accessible to the general mathematician motivated by the applications of the theory to the Poincaré conjecture, and Thurston's geometrisation conjecture (which are discussed briefly in Sections 1.4 and 1.5). This has obviously affected my choice of emphasis. I have suppressed some of the analytical issues, as discussed below, but have compiled a list of relevant Riemannian geometry calculations in Chapter 2.

There are some extremely important aspects of the theory which do not get a mention in these notes. For example, Perelman's $\mathcal{L}$-length, which is a key tool when developing the theory further, and Hamilton's Harnack estimates. There is no discussion of the Kähler-Ricci flow.

I have stopped just short of proving the Hamilton-Ivey pinching result which makes the study of singularities in three-dimensions tractable, although I have covered the necessary techniques to deal with this, and may add an exposition at a later date.

The notes are not completely self-contained. In particular, I state/use the following without giving full proofs:

(i) Existence and uniqueness theory for quasilinear parabolic equations on vector bundles. This is a long story involving rather different techniques to those I focus on in this work. Unfortunately, it is not feasible just to quote theorems from existing sources, and one must learn this theory for oneself;

(ii) Compactness theorems for manifolds and flows. The full proofs of these are long, but a treatment of Ricci flow without using them would be very misleading;

(iii) Parts of Lemma 8.1.8 which involves analysis beyond the level I was assuming. I have given a reference, and intend to give a simple proof in later notes.

These notes are published in the L.M.S. Lecture notes series, in conjunction with Cambridge University Press, and are also available at:

`http://www.maths.warwick.ac.uk/~topping/RFnotes.html`

Readers are invited to send comments and corrections to:

topping@maths.warwick.ac.uk

I would like to thank the audience of the course for making some useful comments, especially Young Choi and Mario Micallef. Thanks also to John Lott for comments on and typographical corrections to a 2005 version of the notes. Parts of the original course benefited from conversations with a number of people, including Klaus Ecker and Miles Simon. Brendan Owens and Gero Friesecke have kindly pointed out some typographical mistakes. Parts of the notes have been prepared whilst visiting the University of Nice, the Albert Einstein Max-Planck Institute in Golm and Free University in Berlin, and I would like to thank these institutions for their hospitality. Finally, I would like to thank Neil Course for preparing all the figures, turning a big chunk of the original course notes into LaTeX, and making some corrections.

1

Introduction

1.1 Ricci flow: what is it, and from where did it come?

Our starting point is a smooth closed (that is, compact and without boundary) manifold $\mathcal{M}$, equipped with a smooth Riemannian metric g. Ricci flow is a means of processing the metric g by allowing it to evolve under the PDE

$$\frac{\partial g}{\partial t} = -2\operatorname{Ric}(g) \qquad (1.1.1)$$

where $\operatorname{Ric}(g)$ is the Ricci curvature.

In simple situations, the flow can be used to deform g into a metric distinguished by its curvature. For example, if $\mathcal{M}$ is two-dimensional, the Ricci flow, once suitably renormalised, deforms a metric conformally to one of constant curvature, and thus gives a proof of the two-dimensional uniformisation theorem – see Sections 1.4 and 1.5. More generally, the topology of $\mathcal{M}$ may preclude the existence of such distinguished metrics, and the Ricci flow can then be expected to develop a singularity in finite time. Nevertheless, the behaviour of the flow may still serve to tell us much about the topology of the underlying manifold. The present strategy is to stop a flow once a singularity has formed, and then carefully perform 'surgery' on the evolved manifold, excising any singular regions before continuing the flow. Provided we understand the structure of finite time singularities sufficiently well, we may hope to keep track of the change in topology of the manifold under surgery, and reconstruct the topology of the original manifold from a limiting flow, together with the singular regions removed. In these notes, we develop some key elements of the machinery used to analyse singularities, and demonstrate their use by proving Hamilton's theorem that closed three-manifolds which admit a metric of positive Ricci curvature also admit a metric of constant positive sectional curvature.

Of all the possible evolutions for g, one may ask why (1.1.1) has been chosen. As we shall see later, in Section 6.1, one might start by considering a gradient flow for the total scalar curvature of the metric g. This leads to an evolution equation

$$\frac{\partial g}{\partial t} = -\text{Ric} + \frac{R}{2}g,$$

where R is the scalar curvature of g. Unfortunately, this turns out to behave badly from a PDE point of view (see Section 6.1) in that we cannot expect the existence of solutions for arbitrary initial data. Ricci flow can be considered a modification of this idea, first considered by Hamilton [19] in 1982. Only recently, in the work of Perelman [31], has the Ricci flow itself been given a gradient flow formulation (see Chapter 6).

Another justification of (1.1.1) is that from certain viewpoints, Ric(g) may be considered as a Laplacian of the metric g, making (1.1.1) a variation on the usual heat equation. For example, if for a given metric g we choose harmonic coordinates $\{x^i\}$, then for each fixed pair of indices i and j, we have

$$R_{ij} = -\frac{1}{2}\Delta g_{ij} + \text{lower order terms}$$

where R_{ij} is the corresponding coefficient of the Ricci tensor, and Δ is the Laplace-Beltrami operator which is being applied to the function g_{ij}. Alternatively, one could pick normal coordinates centred at a point p, and then compute that

$$R_{ij} = -\frac{3}{2}\Delta g_{ij}$$

at p, with Δ again the Laplace-Beltrami operator. Beware here that the notation Δg_{ij} would normally refer to the coefficient $(\Delta g)_{ij}$, where Δ is the connection Laplacian (that is, the 'rough' Laplacian) but Δg is necessarily zero since the metric is parallel with respect to the Levi-Civita connection.

1.2 Examples and special solutions

1.2.1 Einstein manifolds

A simple example of a Ricci flow is that starting from a round sphere. This will evolve by shrinking homothetically to a point in finite time.

More generally, if we take a metric g_0 such that

$$\text{Ric}(g_0) = \lambda g_0$$

for some constant $\lambda \in \mathbb{R}$ (these metrics are known as Einstein metrics) then a solution $g(t)$ of (1.1.1) with $g(0) = g_0$ is given by

$$g(t) = (1 - 2\lambda t)g_0.$$

(It is worth pointing out here that the Ricci tensor is invariant under uniform scalings of the metric.) In particular, for the round 'unit' sphere (S^n, g_0), we have $\text{Ric}(g_0) = (n - 1)g_0$, so the evolution is $g(t) = (1 - 2(n - 1)t)g_0$ and the sphere collapses to a point at time $T = \frac{1}{2(n-1)}$.

An alternative example of this type would be if g_0 were a hyperbolic metric – that is, of constant sectional curvature -1. In this case $\text{Ric}(g_0) = -(n - 1)g_0$, the evolution is $g(t) = (1 + 2(n - 1)t)g_0$ and the manifold *expands* homothetically for all time.

1.2.2 Ricci solitons

There is a more general notion of self-similar solution than the uniformly shrinking or expanding solutions of the previous section. We consider these 'Ricci solitons' without the assumption that $\mathcal{M}$ is compact. To understand such solutions, we must consider the idea of modifying a flow by a family of diffeomorphisms. Let $X(t)$ be a time dependent family of smooth vector fields on $\mathcal{M}$, generating a family of diffeomorphisms ψ_t. In other words, for a smooth $f : \mathcal{M} \to \mathbb{R}$, we have

$$X(\psi_t(y), t)f = \frac{\partial f \circ \psi_t}{\partial t}(y). \qquad (1.2.1)$$

Of course, we could start with a family of diffeomorphisms ψ_t and define $X(t)$ from it, using (1.2.1).

Next, let $\sigma(t)$ be a smooth function of time.

Proposition 1.2.1. *Defining*

$$\hat{g}(t) = \sigma(t)\psi_t^*(g(t)), \qquad (1.2.2)$$

we have

$$\frac{\partial \hat{g}}{\partial t} = \sigma'(t)\psi_t^*(g) + \sigma(t)\psi_t^*\left(\frac{\partial g}{\partial t}\right) + \sigma(t)\psi_t^*(\mathcal{L}_X g). \qquad (1.2.3)$$

This follows from the definition of the Lie derivative. (It may help you to write $\psi_t^*(g(t)) = \psi_t^*(g(t) - g(s)) + \psi_t^*(g(s))$ and differentiate at $t = s$.) As a consequence of this proposition, if we have a metric g_0, a vector field Y and $\lambda \in \mathbb{R}$ (all independent of time) such that

$$-2\text{Ric}(g_0) = \mathcal{L}_Y g_0 - 2\lambda g_0, \qquad (1.2.4)$$

then after setting $g(t) = g_0$ and $\sigma(t) := 1 - 2\lambda t$, if we can integrate the t-dependent vector field $X(t) := \frac{1}{\sigma(t)} Y$, to give a family of diffeomorphisms ψ_t with ψ_0 the identity, then $\hat{g}$ defined by (1.2.2) is a Ricci flow with $\hat{g}(0) = g_0$:

$$
\begin{aligned}
\frac{\partial \hat{g}}{\partial t} &= \sigma'(t)\psi_t^*(g_0) + \sigma(t)\psi_t^*(\mathcal{L}_X g_0) = \psi_t^*(-2\lambda g_0 + \mathcal{L}_Y g_0) \\
&= \psi_t^*(-2\mathrm{Ric}(g_0)) \\
&= -2\mathrm{Ric}(\psi_t^* g_0) \\
&= -2\mathrm{Ric}(\hat{g}).
\end{aligned}
$$

(Note again that the Ricci tensor is invariant under uniform scalings of the metric.)

Definition 1.2.2.   Such a flow is called a steady, expanding or shrinking 'Ricci soliton' depending on whether $\lambda = 0$, $\lambda < 0$ or $\lambda > 0$ respectively.

Conversely, given any Ricci flow $\hat{g}(t)$ of the form (1.2.2) for some $\sigma(t)$, ψ_t, and $g(t) = g_0$, we may differentiate (1.2.2) at $t = 0$ (assuming smoothness) to show that g_0 is a solution of (1.2.4) for appropriate Y and λ. If we are in a situation where we can be sure of uniqueness of solutions (see Theorem 5.2.2 for one such situation) then our $\hat{g}(t)$ must be the Ricci soliton we have recently constructed.[1]

Definition 1.2.3.   A Ricci soliton whose vector field Y can be written as the gradient of some function $f : \mathcal{M} \to \mathbb{R}$ is known as a 'gradient Ricci soliton.'

In this case, we may compute that $\mathcal{L}_Y g_0 = 2\mathrm{Hess}_{g_0}(f)$ (we will review this fact in (2.3.9) below) and so by (1.2.4), f satisfies

$$\mathrm{Hess}_{g_0}(f) + \mathrm{Ric}(g_0) = \lambda g_0. \tag{1.2.5}$$

Hamilton's cigar soliton (a.k.a. Witten's black hole)
Let $\mathcal{M} = \mathbb{R}^2$, and write $g_0 = \rho^2(dx^2 + dy^2)$, using the convention $dx^2 = dx \otimes dx$. The formula for the Gauss curvature is

$$K = -\frac{1}{\rho^2} \Delta \ln \rho,$$

where this time we are writing $\Delta = \frac{\partial^2}{\partial x^2} + \frac{\partial^2}{\partial y^2}$, and the Ricci curvature can be written in terms of the Gauss curvature as $\mathrm{Ric}(g_0) = K g_0$. If now we set

---

[1] One should beware that uniqueness may fail in general. For example, one can have two distinct (smooth) Ricci flows on a time interval $[0, T]$ starting at the same (incomplete) g_0, even if we ask that each is a soliton for $t \in (0, T]$. (See [40].)

$\rho^2 = \frac{1}{1+x^2+y^2}$, then we find that $K = \frac{2}{1+x^2+y^2}$, that is,

$$\text{Ric}(g_0) = \frac{2}{1+x^2+y^2} g_0. \tag{1.2.6}$$

Meanwhile, if we define Y to be the radial vector field $Y := -2(x\frac{\partial}{\partial x} + y\frac{\partial}{\partial y})$, then one can calculate that

$$\mathcal{L}_Y g_0 = -\frac{4}{1+x^2+y^2} g_0.$$

Therefore by (1.2.4), g_0 flows as a steady ($\lambda = 0$) Ricci soliton.

It is illuminating to write g_0 in terms of the geodesic distance from the origin s, and the polar angle θ to give

$$g_0 = ds^2 + \tanh^2 s\, d\theta^2.$$

This shows that the cigar opens at infinity like a cylinder – and therefore looks like a cigar! It is useful to know the curvature in these coordinates:

$$K = \frac{2}{\cosh^2 s}.$$

Finally, note that the cigar is also a *gradient* soliton since Y is radial. Indeed, we may take $f = -2\ln\cosh s$.

The cigar is one of many Ricci solitons which can be written down explicitly. However, it has been distinguished historically as part of one of the possible limits one could find when making an appropriate rescaling (or "blow-up") of *three*-dimensional Ricci flows near finite-time singularities. Only recently, with work of Perelman, has this possibility been ruled out. The blowing-up of flows near singularities will be discussed in Sections 7.3 and 8.5.

The Bryant soliton

There is a similar rotationally symmetric steady gradient soliton for $\mathbb{R}^3$, found by Bryant. Instead of opening like a cylinder at infinity (as is the case for the cigar soliton) the Bryant soliton opens asymptotically like a paraboloid. It has positive sectional curvature.

The Gaussian soliton

One might consider the stationary (that is, time independent) flow of the standard flat metric on $\mathbb{R}^n$ to be quite boring. However, it may later be useful to consider it as a gradient Ricci soliton in more than one way. First, one may take $\lambda = 0$ and $Y \equiv 0$, and see it as a steady soliton. However, for *any* $\lambda \in \mathbb{R}$, one may set $f(x) = \frac{\lambda}{2}|x|^2$, to see the flow as either an expanding or shrinking soliton depending on the sign of λ. (Note that $\psi_t(x) = (1+\lambda t)x$, and $\mathcal{L}_Y g = 2\lambda g$.)

1.2.3 Parabolic rescaling of Ricci flows

Suppose that $g(t)$ is a Ricci flow for $t \in [0, T]$. (Implicit in this statement here, and throughout these notes, is that $g(t)$ is a smooth family of smooth metrics – smooth all the way to $t = 0$ and $t = T$ – which satisfies (1.1.1).) Given a scaling factor $\lambda > 0$, if one defines a new flow by scaling time by λ and distances by $\lambda^{\frac{1}{2}}$, that is one defines

$$\hat{g}(x, t) = \lambda g(x, t/\lambda), \qquad (1.2.7)$$

for $t \in [0, \lambda T]$, then

$$\frac{\partial \hat{g}}{\partial t}(x, t) = \frac{\partial g}{\partial t}(x, t/\lambda) = -2\mathrm{Ric}(g(t/\lambda))(x) = -2\mathrm{Ric}(\hat{g}(t))(x), \qquad (1.2.8)$$

and so $\hat{g}$ is also a Ricci flow. Under this scaling, the Ricci tensor is invariant, as we have just used again, but sectional curvatures and the scalar curvature are scaled by a factor λ^{-1}; for example,

$$R(\hat{g}(x, t)) = \lambda^{-1} R(g(x, t/\lambda)). \qquad (1.2.9)$$

The connection also remains invariant.

The main use of this rescaling will be to analyse Ricci flows which develop singularities. We will see in Section 5.3 that such flows have curvature which blows up (that is, tends to infinity in magnitude) and much of our effort during these notes will be to develop a way of rescaling the flow where the curvature is becoming large in such a way that we can pass to a limit which will be a new Ricci flow encoding some of the information contained in the singularity. This is a very successful strategy in many branches of geometric analysis. Blow-up limits in other problems include tangent cones of minimal surfaces and bubbles in the harmonic map flow.

1.3 Getting a feel for Ricci flow

We have already seen some explicit, rigorous examples of Ricci flows, but it is important to get a feel for how we expect more general Ricci flows, with various shapes and dimensions, to evolve. We approach this from a purely heuristic point of view.

1.3.1 Two dimensions

In two dimensions, we know that the Ricci curvature can be written in terms of the Gauss curvature K as $\mathrm{Ric}(g) = Kg$. Working directly from the equation

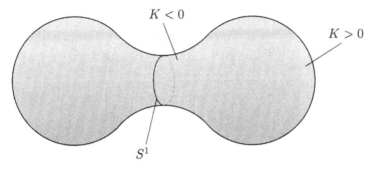

Figure 1.1 2-sphere

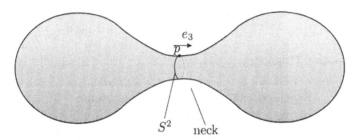

Figure 1.2 3-sphere

(1.1.1), we then see that regions in which $K < 0$ tend to expand, and regions where $K > 0$ tend to shrink.

By inspection of Figure 1.1, one might then guess that the Ricci flow tends to make a 2-sphere "rounder". This is indeed the case, and there is an excellent theory which shows that the Ricci flow on any closed surface tends to make the Gauss curvature constant, after renormalisation. See the book of Chow and Knopf [7] for more information about this specific dimension.

1.3.2 Three dimensions

The neck pinch

The three-dimensional case is more complicated, but we can gain useful intuition by considering the flow of an analogous three-sphere.

Now the cross-sectional sphere is an S^2 (rather than an S^1 as it was before) as indicated in Figure 1.2, and it has its own positive curvature. Let e_1, e_2, e_3 be orthonormal vectors at the point p in Figure 1.2, with e_3 perpendicular to the indicated cross-sectional S^2. Then the sectional curvatures $K_{e_1 \wedge e_3}$ and $K_{e_2 \wedge e_3}$ of the 'planes' $e_1 \wedge e_3$ and $e_2 \wedge e_3$ are slightly negative (c.f. Figure 1.1) but $K_{e_1 \wedge e_2}$

(*i*)

(*ii*)

(*iii*)

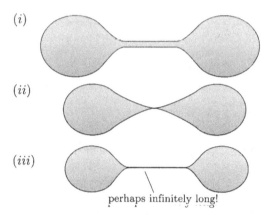

perhaps infinitely long!

Figure 1.3 Neck Pinch

is very positive. Therefore

$$\mathrm{Ric}(e_1, e_1) = K_{e_1 \wedge e_2} + K_{e_1 \wedge e_3} = \text{very positive}$$
$$\mathrm{Ric}(e_2, e_2) = K_{e_2 \wedge e_1} + K_{e_2 \wedge e_3} = \text{very positive}$$
$$\mathrm{Ric}(e_3, e_3) = K_{e_3 \wedge e_1} + K_{e_3 \wedge e_2} = \text{slightly negative}$$

This information indicates how the manifold begins to evolve under the Ricci flow equation (1.1.1). We expect that distances shrink rapidly in the e_1 and e_2 directions, but expand slowly in the e_3 direction. Thus, the metric wants to quickly contract the cross-sectional S^2 indicated in Figure 1.2, whilst slowly stretching the neck. At later times, we expect to see something like the picture in Figure 1.3(*i*) and perhaps eventually 1.3(*ii*) or maybe even 1.3(*iii*).

Only recently have theorems been available which rigorously establish that something like this behaviour does sometimes happen. For more information, see [1] and [37].

It is important to get some understanding of the exact structure of this process. Singularities are typically analysed by blowing up: Where the curvature is large, we magnify – that is, rescale or 'blow-up' – so that the curvature is no longer large, as in Figure 1.4. (Recall the discussion of rescaling in Section 1.2.3.) We will work quite hard to make this blowing-up process precise and rigorous, with the discussion centred on Sections 7.3 and 8.5.

In this particular instance, the blow-up looks like a part of the cylinder $S^2 \times \mathbb{R}$ (a 'neck') and the most advanced theory in three-dimensions tells us that in some sense this is typical. See [31] for more information.

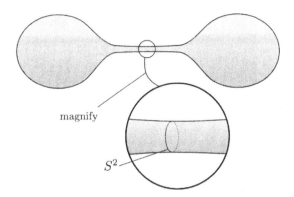

Figure 1.4 Blowing up

The degenerate neck pinch

One possible blow-up, the existence of which we shall not try to make rigorous, is the degenerate neck pinch. Consider the flow of a similar, but asymmetrical three-sphere of the following form:

If the part R is small, then the flow should look like:

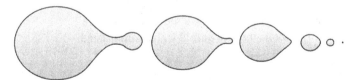

and the manifold should look asymptotically like a small sphere. Meanwhile, if the part R is large, then the flow should look like:

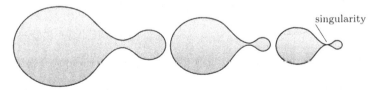

Somewhere in between (when R is of just the right size), we should have:

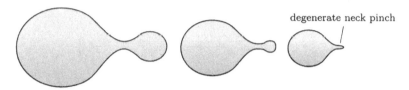

If we were to blow-up this singularity, then we should get something looking like the Bryant soliton:

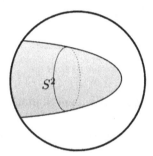

Figure 1.5 Magnified degenerate neck pinch

Infinite time behaviour

A Ricci flow need not converge as $t \to \infty$. In our discussion of Einstein manifolds (Section 1.2.1) we saw that a hyperbolic manifold continues to expand forever, and in Section 1.2.2 we wrote down examples such as the cigar soliton which evolve in a more complicated way. Even if we renormalise our flow, or adjust it by a time-dependent diffeomorphism, we cannot expect convergence, and the behaviour of the flow could be quite complicated. We now give a rough sketch of one flow we should expect to see at 'infinite time'.

Imagine a hyperbolic three-manifold with a toroidal end.

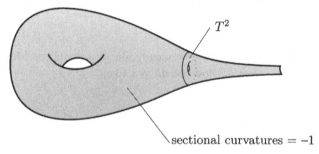

This would expand homothetically under the Ricci flow, as we discussed in Section 1.2.1. Now paste two such pieces together, breaking the hyperbolicity of the metric near the pasting region, and flow:

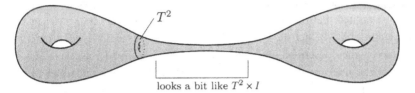

looks a bit like $T^2 \times I$

The ends where the manifold is roughly hyperbolic should tend to expand, but the $T^2 \times I$ 'neck' part should be pretty flat and then wouldn't tend to move much. Much later, the ends should be huge. Scaling down to normalise the volume, the picture should be

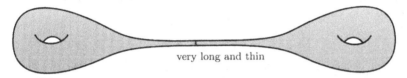

very long and thin

1.4 The topology and geometry of manifolds in low dimensions

Let us consider only closed, oriented manifolds in this section. (Our manifolds are always assumed to be smooth and connected.) One would like to list all such manifolds, and describe them in terms of the geometric structures they support. We now sketch some of what is known on this topic, since Ricci flow turns out to be a useful tool in addressing such problems. This is purely motivational, and the rest of these notes could be read independently.

Dimension 1
There is only one such manifold: the circle S^1, and there is no intrinsic geometry to discuss.

Dimension 2
Such surfaces are classified by their genus $g \in \mathbb{N} \cup \{0\}$. First we have the 2-sphere S^2 ($g = 0$) and second, the torus T^2 ($g = 1$). Then, there are the genus $g \geq 2$ surfaces which arise by taking the *connected sum* of g copies of T^2. (See Appendix A for a description of the notion of connected sum.)

There is an elegant geometric picture lying behind this classification. It turns out that each such surface can be endowed with a conformally equivalent metric of constant Gauss curvature. By uniformly scaling the metric, we may assume that the curvature is $K = 1$, 0 or -1. Once we have this special metric, it can be argued that the universal cover of the surface must be S^2, $\mathbb{R}^2$ or $\mathbb{H}^2$ depending on whether the curvature is 1, 0 or -1 respectively. The original surface is then described, up to conformal change of metric, as a quotient of one of the three model spaces S^2, $\mathbb{R}^2$ or $\mathbb{H}^2$ by a discrete subgroup of the group of isometries, acting freely. This gives rise to S^2, a flat torus, or a higher genus hyperbolic surface depending on whether the curvature is 1, 0 or -1 respectively. In particular, we have the Uniformisation Theorem describing all compact Riemann surfaces.

To help draw analogies with the three-dimensional case, let us note that $\pi_1(S^2) = 1$, $\pi_1(T^2) = \mathbb{Z} \oplus \mathbb{Z}$ and for the higher genus surfaces, π_1 is infinite, but does not contain $\mathbb{Z} \oplus \mathbb{Z}$ as a subgroup.

Dimension 3

Several decades ago, Thurston conjectured a classification for the three-dimensional case which has some parallels with the two-dimensional case. In this section we outline some of the main points of this story. Further information may be obtained from [38], [35], [24], [26] and [13].

One formulation of the conjecture (see [13]) analogous to the two dimensional theory, states that if our manifold $\mathcal{M}$ is also *irreducible* (which means that every 2-sphere embedded in the manifold bounds a three-ball) then precisely one of the following holds:

(i) $\mathcal{M} = S^3/\Gamma$, with $\Gamma \subset Isom(S^3)$;
(ii) $\mathbb{Z} \oplus \mathbb{Z} \subset \pi_1(\mathcal{M})$;
(iii) $\mathcal{M} = \mathbb{H}^3/\Gamma$, with $\Gamma \subset Isom(\mathbb{H}^3)$.

Case (ii) holds if $\mathcal{M}$ contains an incompressible torus; in other words, if there exists an embedding $\phi : T^2 \to \mathcal{M}$ for which the induced map $\pi_1(T^2) \to \pi_1(\mathcal{M})$ is injective. A (nontrivial) partial converse is that case (ii) implies that either $\mathcal{M}$ contains an incompressible torus, or $\mathcal{M}$ is a so-called Seifert fibred space – see [35].

If our manifold is not irreducible, then we may first have to perform a decomposition. We say that a three-manifold is *prime* if it cannot be expressed as a nontrivial connected sum of two other manifolds. (A *trivial* connected sum decomposition would be to write a manifold as the sum of itself with S^3.) One can show (see [24]) that any prime three-manifold (orientable etc. as throughout Section 1.4) is either irreducible or $S^1 \times S^2$.

The classical theorem of Kneser (1928) tells us that any of our manifolds may be decomposed into a connected sum of finitely many prime manifolds – see

[24]. At this point we may address the irreducible components with Thurston's conjecture as stated above.

Although the conjecture as stated above looks superficially like its two-dimensional analogue, the case (ii) lacks the geometric picture that we had before. Note that manifolds in this case cannot in general be equipped with a metric of constant sectional curvature. For example, the product of a hyperbolic surface and S^1 does not support such a metric.

Instead, we try to write all of our prime manifolds as compositions of 'geometric' pieces in the sense of the second definition below. Manifolds within case (ii) above may require further decomposition before we can endow them with one of several geometrically natural metrics.

Definition 1.4.1. A *geometry* is a simply-connected homogeneous unimodular Riemannian manifold X.

Here, homogeneous means that given any two points in the manifold $\mathcal{M}$, there exists an isometry $\mathcal{M} \mapsto \mathcal{M}$ mapping one point to the other. Unimodular means that X admits a discrete group of isometries with compact quotient.

These 'geometries' can be classified in three dimensions. There are eight of them:

- S^3, $\mathbb{R}^3$, $\mathbb{H}^3$ – the constant curvature geometries,
- $S^2 \times \mathbb{R}$, $\mathbb{H}^2 \times \mathbb{R}$ – the product geometries,
- Nil, Sol, $\widetilde{SL_2}(\mathbb{R})$ – are twisted product geometries.

See [35] or [38] for a description of the final three of these, and a proof that these are the only geometries.

Definition 1.4.2. A compact manifold $\mathcal{M}$ (possibly with boundary) is called *geometric* if $int(\mathcal{M}) = X/\Gamma$ has finite volume, where X is a geometry and Γ is a discrete group of isometries acting freely.

When a geometric manifold does have a boundary, it can only consist of a union of incompressible tori. (This can only occur for quotients of the geometries $\mathbb{H}^3$, $\mathbb{H}^2 \times \mathbb{R}$ and $\widetilde{SL_2}(\mathbb{R})$.)

Conjecture 1.4.3 (Thurston's geometrisation conjecture). Any (smooth, closed, oriented) prime three-manifold arises by gluing a finite number of 'geometric' pieces along their boundary tori.

In Chapter 10, we will prove a special case of this conjecture, due to Hamilton, using the Ricci flow. In the case that our manifold admits a metric of positive Ricci curvature (which forces its fundamental group to be finite by Myer's theorem [27, Theorem 11.8]) we will show that we indeed lie in case (i) – that is, our manifold is a quotient of S^3 – by showing that the manifold carries a metric of constant positive sectional curvature, and therefore its universal cover must be S^3.

Dimension 4
By now the problem·is much harder, and one hopes only to classify such manifolds under some extra hypothesis – for instance a curvature constraint.

1.5 Using Ricci flow to prove topological and geometric results

Dimension 2
In two dimensions, the Ricci flow, once suitably renormalised, flows arbitrary metrics to metrics of constant curvature, and remains in the same conformal class.

Very recently (see [6]) the proof of this fact has been adjusted to remove any reliance on the Uniformisation theorem, and so by finding this special constant curvature conformal metric, the Ricci flow itself proves the Uniformisation Theorem for compact Riemann surfaces, as discussed in Section 1.4. See [7] for more information about the two-dimensional theory prior to [6].

Dimension 3
There is a strategy for proving Thurston's geometrisation conjecture, due to Hamilton (1980s and 1990s) based partly on suggestions of Yau, which has received a boost from work of Perelman, since 2002. In this section we aim only to give a heuristic outline of this programme.

The idea is to start with an arbitrary metric on $\mathcal{M}$ and flow. Typically, we would expect to see singularities like neck pinches:

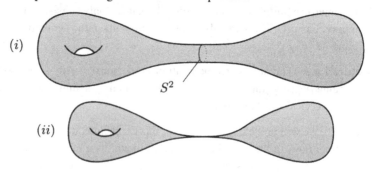

Figure 1.6 Neck pinch

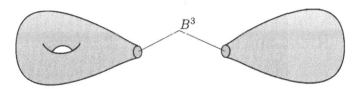

Figure 1.7 Surgery

At the singular time, we chop out the neck and paste in B^3 caps (see Figure 1.7) and then restart the flow for each component.

Heuristically, one hopes this procedure is performing the prime decomposition. Unfortunately this 'surgery' procedure might continue forever, involving infinitely many surgeries, but Perelman has claimed that there are only finitely many surgeries required over any finite time interval when the procedure is done correctly. See [32] for more details.

In some cases, for example when the manifold is simply connected, all the flow eventually disappears – see [33] and [8] – and this is enough to establish the Poincaré Conjecture, modulo the details of the surgery procedure.

In general, when the flow does not become extinct, one would like to show that the metric at large times is sufficiently special that we can understand the topology of the underlying manifold via a so-called 'thick-thin' decomposition. See [32] for more details.

The only case we shall cover in detail in these notes is that in which the Ricci curvature of the initial metric is positive. The flow of such a metric is analogous to the flow of an arbitrary metric on S^2 in that it converges (once suitably renormalised) to a metric of constant positive sectional curvature, without requiring surgery, as we show in Chapter 10.

Dimension 4

Ricci flow has had some success in describing four-manifolds of positive isotropic curvature. See the paper [23] of Hamilton, to which corrections are required.

2
Riemannian geometry background

2.1 Notation and conventions

Throughout this chapter, X, Y, W and Z will be fixed vector fields on a manifold $\mathcal{M}$, and A and B will be more general tensor fields, with possibly different type.[1]

We assume that $\mathcal{M}$ is endowed with a Riemannian metric g, or a smooth family of such metrics depending on one parameter t. This metric then extends to arbitrary tensors. (See, for example, [27, Lemma 3.1].)

We will write ∇ for the Levi-Civita connection of g. Recall that it may be extended to act on arbitrary tensor fields. (See, for example [27, Lemma 4.6].) It is also important to keep in mind that the Levi-Civita connection commutes with traces, or equivalently that $\nabla g = 0$, and that we have the product rule $\nabla_X(A \otimes B) = (\nabla_X A) \otimes B + A \otimes (\nabla_X B)$. See [27, Lemma 4.6] for more on what this implies. When we apply ∇ without a subscript, we adopt the convention of seeing $\nabla_X A = \nabla A(X, \ldots)$, that is, the X appears as the first rather than the last entry. (More details, could be found in [14, (2.60)] or with the opposite convention in [27, Lemma 4.7].)

We need notation for the second covariant derivative of a tensor field, and write

$$\nabla^2_{X,Y} := \nabla_X \nabla_Y - \nabla_{\nabla_X Y}.$$

That way, we have

$$\nabla^2_{X,Y} A := \nabla_X \nabla_Y A - \nabla_{\nabla_X Y} A \equiv (\nabla^2 A)(X, Y, \ldots),$$

where $\nabla^2 A$ is the covariant derivative of ∇A. When applied to a function, we get the Hessian

$$\mathrm{Hess}(f) := \nabla df.$$

---

[1] A tensor field of type (p, q), with $p, q \in \{0\} \cup \mathbb{N}$ is a section of the tensor product of the bundles $\otimes^p T\mathcal{M}$ and $\otimes^q T^*\mathcal{M}$.

We adopt the sign convention

$$\Delta A := \operatorname{tr}_{12} \nabla^2 A,$$

for the connection (or rough) Laplacian of A, where tr_{12} means to trace over the first and second entries (here of $\nabla^2 A$).

We adopt the sign convention

$$\begin{aligned} R(X, Y) &:= \nabla_Y \nabla_X - \nabla_X \nabla_Y + \nabla_{[X,Y]} \\ &= \nabla^2_{Y,X} - \nabla^2_{X,Y}. \end{aligned} \tag{2.1.1}$$

for the curvature. This may be applied to any tensor field, and again satisfies the product rule $R(X, Y)(A \otimes B) = (R(X, Y)A) \otimes B + A \otimes (R(X, Y)B)$. Of course, $R(X, Y)f = 0$ for any function $f : \mathcal{M} \to \mathbb{R}$ – equivalently the Hessian Hess(f) of a function f is symmetric – but $\nabla^2_{X,Y}$ is not otherwise symmetric in X and Y. Indeed, we have, for any tensor field $A \in \Gamma(\otimes^k T^* \mathcal{M})$, the Ricci identity

$$\begin{aligned} &-\nabla^2_{X,Y} A(W, Z, \ldots) + \nabla^2_{Y,X} A(W, Z, \ldots) \\ &\quad \equiv [R(X, Y)A](W, Z, \ldots) \\ &\quad \equiv [R(X, Y)A](W, Z, \ldots) - R(X, Y)[A(W, Z, \ldots)] \\ &\quad \equiv -A(R(X, Y)W, Z, \ldots) - A(W, R(X, Y)Z, \ldots) - \ldots. \end{aligned} \tag{2.1.2}$$

We also may write[2]

$$\operatorname{Rm}(X, Y, W, Z) := \langle R(X, Y)W, Z \rangle.$$

Note that about half – perhaps more – of books adopt the opposite sign convention for $R(X, Y)$. We agree with [10], [14], [29], etc. but not with [27] etc. This way makes more sense because we then have the agreement with classical notation:

$$\operatorname{Rm}(\partial_i, \partial_j, \partial_k, \partial_l) = R_{ijkl}.$$

(A few people adopt the opposite sign convention for R_{ijkl}.) This way round, $R(X, Y)$ then serves, roughly speaking, as the infinitesimal *holonomy* rotation as one parallel translates around an infinitesimal *anticlockwise* loop in the 'plane' $X \wedge Y$. It also leads to a more natural definition of the curvature operator $\mathcal{R} : \wedge^2 T \mathcal{M} \to \wedge^2 T \mathcal{M}$, namely

$$\langle \mathcal{R}(X \wedge Y), W \wedge Z \rangle = \operatorname{Rm}(X, Y, W, Z). \tag{2.1.3}$$

---

[2] We use $g(\cdot, \cdot)$ and $\langle \cdot, \cdot \rangle$ interchangeably, although with the latter, it is easier to forget any t-dependence that g might have.

The various symmetries of Rm ensure that $\mathcal{R}$ is thus well defined, and is symmetric. If X and Y are orthogonal unit vectors at some point, then the sectional curvature of the plane $X \wedge Y$ is $Rm(X, Y, X, Y)$ with our convention.

We denote the Ricci and scalar curvatures by

$$Ric(X, Y) := \operatorname{tr} Rm(X, \cdot, Y, \cdot),$$

and

$$R := \operatorname{tr} Ric,$$

respectively. (There should not be any ambiguity about the sign of these!) Occasionally we write, say, $Ric(g)$ or Ric_g to emphasise the metric we are using.

Some of the formulae which we will be requiring can be dramatically simplified, without losing relevant information, by using the following '$*$'-notation. We denote by $A * B$ any tensor field which is a (real) linear combination of tensor fields, each formed by starting with the tensor field $A \otimes B$, using the metric to switch the type of any number of $T^*\mathcal{M}$ components to $T\mathcal{M}$ components, or vice versa (that is, raising or lowering some indices) taking any number of contractions, and switching any number of components in the product. Here, the algorithm for arriving at a certain expression $A * B$ must be independent of the particular choice of tensors A and B of their respective types, and hence we are free to estimate $|A * B| \leq C|A||B|$ for some constant C which will depend neither on A nor B. Generally, the precise dependencies of C should be clear in context, and in these notes we will typically have $C = C(n)$ when making such an estimate.

The fact that g is parallel gives us the product rule

$$\nabla(A * B) = (\nabla A) * B + A * (\nabla B). \tag{2.1.4}$$

As an example of the use of this notation, the expression $R^2 = Rm * Rm$, or even $R = Rm * 1$, would be valid, albeit not very useful. The Ricci identity (2.1.2) can be weakened to the useful notational aid

$$R(\cdot, \cdot)A = A * Rm, \tag{2.1.5}$$

valid in this form for tensor fields A of arbitrary type. One identity which follows easily from this expression is the formula for commuting the covariant derivative and the Laplacian:

$$\nabla(\Delta A) - \Delta(\nabla A) = \nabla Rm * A + Rm * \nabla A. \tag{2.1.6}$$

We will be regularly using the divergence operator $\delta : \Gamma(\otimes^k T^*\mathcal{M}) \to \Gamma(\otimes^{k-1} T^*\mathcal{M})$ defined by $\delta(T) = -\mathrm{tr}_{12}\nabla T$. Again, tr_{12} means to trace over the first and second entries (here of ∇T).

Remark 2.1.1. The formal adjoint of δ acting on this space of sections is the covariant derivative $\nabla : \Gamma(\otimes^{k-1} T^*\mathcal{M}) \to \Gamma(\otimes^k T^*\mathcal{M})$. However, if one restricts δ to a map from $\Gamma(\wedge^k T^*\mathcal{M})$ to $\Gamma(\wedge^{k-1} T^*\mathcal{M})$ then its formal adjoint is the exterior derivative (up to a constant depending on one's choice of inner product). Moreover, we shall see later that if $k = 2$ and one restricts δ to a map from $\Gamma(\mathrm{Sym}^2 T^*\mathcal{M})$ to $\Gamma(T^*\mathcal{M})$, then its formal adjoint is $\omega \mapsto \frac{1}{2}\mathcal{L}_{\omega^\sharp} g$ where $\sharp$ represents the musical isomorphism $\Gamma(T^*\mathcal{M}) \mapsto \Gamma(T\mathcal{M})$ (see [14, (2.66)], for example).

For various $T \in \Gamma(\mathrm{Sym}^2 T^*\mathcal{M})$ we will need the 'gravitation tensor'

$$G(T) := T - \frac{1}{2}(\mathrm{tr}T)g, \tag{2.1.7}$$

and its divergence

$$\delta G(T) = \delta T + \frac{1}{2}d(\mathrm{tr}T). \tag{2.1.8}$$

A useful identity involving the quantities we have just defined is

$$\delta G(\mathrm{Ric}) = \delta\mathrm{Ric} + \frac{1}{2}dR = 0, \tag{2.1.9}$$

which arises by contracting the second Bianchi identity (see [14, (3.135)]).

2.2 Einstein metrics

As mentioned in Section 1.2.1, an Einstein metric g is one for which there exists a $\lambda \in \mathbb{R}$ for which $\mathrm{Ric}(g) = \lambda g$. Some authors allow λ to be a function $\lambda : \mathcal{M} \to \mathbb{R}$, or equivalently (by taking the trace) ask simply that $\mathrm{Ric}(g) = \frac{R}{n}g$. In this case, one can apply the divergence operator, and use (2.1.9) to find that

$$dR \equiv -\delta(Rg) = -n\delta(\mathrm{Ric}) \equiv \frac{n}{2}dR.$$

Therefore, in dimensions $n \neq 2$, the scalar curvature is constant, and the two definitions agree.

2.3 Deformation of geometric quantities as the Riemannian metric is deformed

Suppose we have a smooth family of metrics $g = g_t \in \Gamma(\mathrm{Sym}^2 T^* \mathcal{M})$ for t in an open interval, and write $h := \frac{\partial g_t}{\partial t}$. We wish to calculate the variation of the curvature tensor, volume form and other geometric quantities as the metric varies, in terms of h and g. These formulae will be used to *linearise* equations such as that of the Ricci flow during our discussion of short-time existence in Chapter 5, to calculate the gradients of functionals such as the total scalar curvature in Chapter 6, and to calculate how the curvature and geometric quantities evolve during the Ricci flow in Section 2.5. Only in the final case will t represent time. First we list the formulae; their derivations will be compiled in Section 2.3.2.

2.3.1 The formulae

First we want to see how the Levi-Civita connection changes as the metric changes.

Proposition 2.3.1.

$$\left\langle \frac{\partial}{\partial t} \nabla_X Y, Z \right\rangle = \frac{1}{2} \left[(\nabla_Y h)(X, Z) + (\nabla_X h)(Y, Z) - (\nabla_Z h)(X, Y) \right].$$

Note in the above that although $\nabla_X Y$ is not a tensor (because $\nabla_X f Y \neq f \nabla_X Y$ for a general function $f : \mathcal{M} \to \mathbb{R}$) you can check that

$$\Pi(X, Y) := \frac{\partial}{\partial t} \nabla_X Y$$

is in fact a tensor. (In alternative language, the Christoffel symbols do not represent a tensor, but their derivatives with respect to t do.)

Remark 2.3.2. If $V = V(t)$ were a t-dependent vector field, we would have instead

$$\frac{\partial}{\partial t} \nabla_X V = \Pi(X, V) + \nabla_X \frac{\partial V}{\partial t}. \tag{2.3.1}$$

Remark 2.3.3. A weakened form of this proposition which suffices for many purposes is

$$\frac{\partial}{\partial t} \nabla Y = Y * \nabla h,$$

where the $*$-notation is from Section 2.1. Moreover, if $\omega \in \Gamma(T^* \mathcal{M})$ is a one-form which is independent of t, then

$$\left(\frac{\partial}{\partial t}\nabla_X \omega\right)(Y) = \frac{\partial}{\partial t}\nabla_X[\omega(Y)] - \frac{\partial}{\partial t}\omega(\nabla_X Y)$$

$$= \frac{\partial}{\partial t}(X[\omega(Y)]) - \omega\left(\frac{\partial}{\partial t}\nabla_X Y\right) \qquad (2.3.2)$$

$$= -\omega\left(\frac{\partial}{\partial t}\nabla_X Y\right),$$

and so

$$\frac{\partial}{\partial t}\nabla\omega = \omega * \nabla h.$$

By the product rule, we then find the formula

$$\frac{\partial}{\partial t}\nabla A = A * \nabla h,$$

for any fixed tensor field A, or more generally, if $A = A(t)$ is given a t-dependency, then we have

$$\frac{\partial}{\partial t}\nabla A - \nabla\frac{\partial}{\partial t}A = A * \nabla h. \qquad (2.3.3)$$

This formula may be compared with (2.1.6).

Next we take a first look at how the curvature is changing.

Proposition 2.3.4.

$$\frac{\partial}{\partial t}R(X, Y)W = (\nabla_Y \Pi)(X, W) - (\nabla_X \Pi)(Y, W)$$

We want to turn this into a formula for the evolution of the full curvature tensor Rm in terms only of $h := \frac{\partial g}{\partial t}$.

Proposition 2.3.5.

$$\frac{\partial}{\partial t}\mathrm{Rm}(X, Y, W, Z) = \frac{1}{2}[h(R(X, Y)W, Z) - h(R(X, Y)Z, W)]$$

$$+ \frac{1}{2}\left[\nabla^2_{Y,W}h(X, Z) - \nabla^2_{X,W}h(Y, Z)\right. \qquad (2.3.4)$$

$$\left. + \nabla^2_{X,Z}h(Y, W) - \nabla^2_{Y,Z}h(X, W)\right]$$

Note that the anti-symmetry between X and Y, and also between W and Z, is automatic in this expression. The tensor Rm also enjoys the symmetry $\mathrm{Rm}(X, Y, W, Z) = \mathrm{Rm}(W, Z, X, Y)$ (that is, the curvature operator is symmetric) and we see this in the right-hand side of (2.3.4) via the Ricci identity

(2.1.2) which in this case tells us that

$$-\nabla^2_{X,Y}h(W, Z) + \nabla^2_{Y,X}h(W, Z) = -h(R(X, Y)W, Z) - h(R(X, Y)Z, W).$$
(2.3.5)

Next we want to compute the evolution of the Ricci and scalar curvatures. Since these arise as traces, we record now the following useful fact.

Proposition 2.3.6. *For any t-dependent tensor $\alpha \in \Gamma(\otimes^2 T^*\mathcal{M})$, there holds*

$$\frac{\partial}{\partial t}(\mathrm{tr}\,\alpha) = -\langle h, \alpha \rangle + \mathrm{tr}\frac{\partial \alpha}{\partial t}.$$

Proposition 2.3.7.

$$\frac{\partial}{\partial t}\mathrm{Ric} = -\frac{1}{2}\Delta_L h - \frac{1}{2}\mathcal{L}_{(\delta G(h))^\#}g,$$
(2.3.6)

where Δ_L is the Lichnerowicz Laplacian

$$(\Delta_L h)(X, W) := (\Delta h)(X, W) - h(X, \mathrm{Ric}(W)) - h(W, \mathrm{Ric}(X))$$
$$+ 2\mathrm{tr}\,h(R(X, \cdot)W, \cdot).$$
(2.3.7)

Remark 2.3.8. In the definition of the Lichnerowicz Laplacian above, we have viewed the Ricci tensor as a section of $T^*\mathcal{M} \otimes T\mathcal{M}$ (using the metric). On such occasions, we will tend to write $\mathrm{Ric}(X)$ for the vector field defined in terms of the usual $\mathrm{Ric}(\cdot, \cdot)$ by $\mathrm{Ric}(X, Y) = \langle \mathrm{Ric}(X), Y \rangle$, or equivalently $\mathrm{Ric}(X) := (\mathrm{Ric}(X, \cdot))^\#$.

A term $\mathcal{L}_X g$ can be viewed as a 'symmetrized gradient' of X – see [14, (2.62)] – and for any 1-form ω we have

$$\mathcal{L}_{\omega^\#}g(X, W) = \nabla\omega(X, W) + \nabla\omega(W, X).$$
(2.3.8)

In the special case that $\omega = df$ for some function $f : \mathcal{M} \to \mathbb{R}$, we then have

$$\mathcal{L}_{(df)^\#}g = \mathcal{L}_{(\nabla f)}g = 2\mathrm{Hess}(f)$$
(2.3.9)

where $\mathrm{Hess}(f)$ is the Hessian – the symmetric tensor ∇df – as before. Combining with (2.1.8) we can expand the final term of (2.3.6) as

$$\mathcal{L}_{(\delta G(h))^\#}g = \mathcal{L}_{(\delta h)^\#}g + \mathrm{Hess}(\mathrm{tr}h).$$
(2.3.10)

It will be important during the proof[3] to be able to juggle different formulations of the lower order terms in the definition of Δ_L. First, we have

$$h(X, \mathrm{Ric}(W)) = \langle h(X, \cdot), \mathrm{Ric}(W, \cdot) \rangle = \mathrm{tr}\,h(X, \cdot) \otimes \mathrm{Ric}(W, \cdot)$$
$$= -\mathrm{tr}\,h(R(W, \cdot)\cdot, X),$$
(2.3.11)

---

[3] In fact, you might like to do this calculation using index notation.

which one can easily check with respect to an orthonormal frame $\{e_i\}$; for example

$$\operatorname{tr} h(X, \cdot) \otimes \operatorname{Ric}(W, \cdot) = \sum_i h(X, e_i)\operatorname{Ric}(W, e_i) = h(X, \operatorname{Ric}(W, e_i)e_i)$$

$$= h(X, \operatorname{Ric}(W)).$$

Similarly, we have

$$\operatorname{tr} h(R(X, \cdot)W, \cdot) = \langle \operatorname{Rm}(X, \cdot, W, \cdot), h \rangle. \tag{2.3.12}$$

There is also an elegant expression for the evolution of the scalar curvature:

Proposition 2.3.9.

$$\frac{\partial}{\partial t} R = -\langle \operatorname{Ric}, h \rangle + \delta^2 h - \Delta(\operatorname{tr} h).$$

Later, we will have cause to take the t-derivative of the divergence of a 1-form; perhaps of the Laplacian of a function. Since the divergence operator depends on the metric, we must take care to use the following:

Proposition 2.3.10. *For any t-dependent 1-form $\omega \in \Gamma(T^*\mathcal{M})$, there holds*

$$\frac{\partial}{\partial t}\delta\omega = \delta\frac{\partial\omega}{\partial t} + \langle h, \nabla\omega \rangle - \langle \delta G(h), \omega \rangle.$$

During our considerations of short-time existence for the Ricci flow, we will even need the t derivative of the divergence of a symmetric 2-tensor:

Proposition 2.3.11. *Suppose $T \in \Gamma(\operatorname{Sym}^2 T^*\mathcal{M})$ is independent of t. Then*

$$\left(\frac{\partial}{\partial t}\delta G(T)\right) Z = -T((\delta G(h))^\#, Z) + \left\langle h, \nabla T(\cdot, \cdot, Z) - \frac{1}{2}\nabla_Z T \right\rangle.$$

Finally, we record how the volume form $dV := *1 \equiv \sqrt{det(g_{ij})}dx^1 \wedge \ldots \wedge dx^n$ evolves as the metric is deformed.

Proposition 2.3.12.

$$\frac{\partial}{\partial t}dV = \frac{1}{2}(\operatorname{tr} h)dV.$$

2.3.2 The calculations

We will start off the proofs working with arbitrary vector fields, to make the calculations more illuminating for beginners. Later, we will exploit the fact that we're dealing with tensors – and only need check the identities at one point – by working with vector fields X, Y etc. which arise from coordinates (and thus

$[X, Y] = 0$) and whose covariant derivatives vanish (that is, $\nabla X = 0$ etc.) at the point in question.

Proof. (Proposition 2.3.1.)   We compute

$$\langle \Pi(X, Y), Z \rangle = \left\langle \frac{\partial}{\partial t} \nabla_X Y, Z \right\rangle = \frac{\partial}{\partial t} g(\nabla_X Y, Z) - h(\nabla_X Y, Z)$$

$$= \frac{\partial}{\partial t} [Xg(Y, Z) - g(Y, \nabla_X Z)] - h(\nabla_X Y, Z)$$

$$= \left[Xh(Y, Z) - h(Y, \nabla_X Z) - g\left(Y, \frac{\partial}{\partial t} \nabla_X Z\right) \right] - h(\nabla_X Y, Z)$$

$$= (\nabla_X h)(Y, Z) - \langle \Pi(X, Z), Y \rangle$$

$$= (\nabla_X h)(Y, Z) - \langle \Pi(Z, X), Y \rangle.$$

Iterating this identity with the X, Y and Z cycled gives

$$\langle \Pi(X, Y), Z \rangle = (\nabla_X h)(Y, Z) - [(\nabla_Z h)(X, Y) - \langle \Pi(Y, Z), X \rangle].$$

Repeating once more yields

$$\langle \Pi(X, Y), Z \rangle = (\nabla_X h)(Y, Z) - (\nabla_Z h)(X, Y) + [(\nabla_Y h)(Z, X) - \langle \Pi(X, Y), Z \rangle],$$

which is what we want. Alternatively, one could differentiate a standard formula like

$$2\langle \nabla_X Y, Z \rangle = X\langle Y, Z \rangle + Y\langle Z, X \rangle - Z\langle X, Y \rangle$$
$$- \langle Y, [X, Z] \rangle - \langle Z, [Y, X] \rangle + \langle X, [Z, Y] \rangle,$$

from, say, [27, (5.1)].   □

Proof. (Proposition 2.3.4.)   By definition,

$$R(X, Y)W := \nabla_Y \nabla_X W - \nabla_X \nabla_Y W + \nabla_{[X,Y]} W,$$

and so

$$\frac{\partial}{\partial t} R(X, Y)W = [\Pi(Y, \nabla_X W) + \nabla_Y(\Pi(X, W))]$$

$$- [\Pi(X, \nabla_Y W) + \nabla_X(\Pi(Y, W))] + \Pi([X, Y], W)$$

$$= (\nabla_Y \Pi)(X, W) - (\nabla_X \Pi)(Y, W)$$

$$+ \Pi(\nabla_Y X - \nabla_X Y + [X, Y], W)$$

$$= (\nabla_Y \Pi)(X, W) - (\nabla_X \Pi)(Y, W)$$

since ∇ is torsion free.   □

Proof. (Proposition 2.3.5.)   To shorten the calculation, we will check this tensor identity at one point $p \in \mathcal{M}$ at which we may assume, without loss of generality, that at a 'time' t of your choosing,

$$\nabla X = \nabla Y = \nabla W = \nabla Z = 0. \qquad (2.3.13)$$

To begin with, since the metric is evolving, we have

$$\frac{\partial}{\partial t}\langle R(X, Y)W, Z\rangle = h(R(X, Y)W, Z) + \left\langle \frac{\partial}{\partial t}R(X, Y)W, Z\right\rangle$$

$$= h(R(X, Y)W, Z) + \langle (\nabla_Y \Pi)(X, W) - (\nabla_X \Pi)(Y, W), Z\rangle$$

by Proposition 2.3.4. Meanwhile, by Proposition 2.3.1 and (2.3.13) we have

$$\langle (\nabla_Y \Pi)(X, W), Z\rangle = Y\langle \Pi(X, W), Z\rangle$$

$$= \frac{1}{2}Y\left[(\nabla_W h)(X, Z) + (\nabla_X h)(W, Z) - (\nabla_Z h)(X, W)\right]$$

$$= \frac{1}{2}\left[(\nabla_Y \nabla_W h)(X, Z) + (\nabla_Y \nabla_X h)(W, Z)\right.$$

$$\left. - (\nabla_Y \nabla_Z h)(X, W)\right]$$

$$= \frac{1}{2}\left[\nabla^2_{Y,W}h(X, Z) + \nabla^2_{Y,X}h(W, Z) - \nabla^2_{Y,Z}h(X, W)\right],$$

and hence

$$\frac{\partial}{\partial t}\langle R(X, Y), W, Z\rangle = h(R(X, Y)W, Z) + \frac{1}{2}\left[\nabla^2_{Y,W}h(X, Z) - \nabla^2_{X,W}h(Y, Z)\right.$$

$$+ \nabla^2_{Y,X}h(W, Z) - \nabla^2_{X,Y}h(W, Z)$$

$$\left. - \nabla^2_{Y,Z}h(X, W) + \nabla^2_{X,Z}h(Y, W)\right].$$

We conclude by using (2.3.5). $\qquad \square$

Proof. (Proposition 2.3.6.) Writing $\alpha = \alpha_{ij}dx^i \otimes dx^j$, and noting that because $\frac{\partial}{\partial t}g_{ij} = h_{ij}$, we have

$$\frac{\partial}{\partial t}g^{ij} = -h^{ij} := -h_{kl}g^{ik}g^{jl},$$

we compute

$$\frac{\partial}{\partial t}(\text{tr}\alpha) = \frac{\partial}{\partial t}(g^{ij}\alpha_{ij}) = -h^{ij}\alpha_{ij} + g^{ij}\frac{\partial \alpha_{ij}}{\partial t} = -\langle h, \alpha\rangle + \text{tr}\frac{\partial \alpha}{\partial t}.$$

$$\square$$

Proof. (Proposition 2.3.7.) First, note that by Proposition 2.3.6,

$$\frac{\partial}{\partial t}\text{Ric}(X, W) = -\langle \text{Rm}(X, \cdot, W, \cdot), h\rangle + \text{tr}\left[\frac{\partial}{\partial t}\text{Rm}(X, \cdot, W, \cdot)\right]. \qquad (2.3.14)$$

Using Proposition 2.3.5 and (2.3.5):

$$\frac{\partial}{\partial t} \mathrm{Rm}(X, Y, W, Z) = \frac{1}{2} \left[h(R(X, Y)W, Z) - h(R(X, Y)Z, W) \right]$$
$$+ \frac{1}{2} \left[\nabla_{Y,W}^2 h(X, Z) - \nabla_{X,W}^2 h(Y, Z) \right.$$
$$\left. + \nabla_{X,Z}^2 h(Y, W) - \nabla_{Y,Z}^2 h(X, W) \right]$$
$$= \frac{1}{2} \left[h(R(X, Y)W, Z) - h(R(X, Y)Z, W) \right.$$
$$\left. + h(R(Y, W)X, Z) + h(R(Y, W)Z, X) \right]$$
$$+ \frac{1}{2} \left[\nabla_{W,Y}^2 h(X, Z) - \nabla_{X,W}^2 h(Y, Z) \right.$$
$$\left. + \nabla_{X,Z}^2 h(Y, W) - \nabla_{Y,Z}^2 h(X, W) \right].$$

In anticipation of tracing this expression over Y and Z (which will happen when we use (2.3.14)) we take a look at the traces of the final four terms above. For the third of these, we have

$$\mathrm{tr}\, \nabla_{X,\cdot}^2 h(\cdot, W) = -(\nabla \delta h)(X, W).$$

Since h is symmetric, the first term is similar:

$$\mathrm{tr}\, \nabla_{W,\cdot}^2 h(X, \cdot) = -(\nabla \delta h)(W, X).$$

For the second we have

$$\mathrm{tr}\, \nabla_{X,W}^2 h(\cdot, \cdot) = \nabla_{X,W}^2 (\mathrm{tr} h) = \mathrm{Hess}(\mathrm{tr} h)(X, W).$$

For the fourth and final term we need only recall the definition of the connection Laplacian,

$$\mathrm{tr}\, \nabla_{\cdot,\cdot}^2 h(X, W) = (\Delta h)(X, W).$$

Combining these four with (2.3.14) and (2.3.12), we find that

$$\frac{\partial}{\partial t} \mathrm{Ric}(X, W) = -\mathrm{tr} \left[h(R(X, \cdot)W, \cdot) \right] + \mathrm{tr} \left[\frac{\partial}{\partial t} \mathrm{Rm}(X, \cdot, W, \cdot) \right]$$
$$= \frac{1}{2} \mathrm{tr} \left[-h(R(X, \cdot)W, \cdot) - h(R(X, \cdot)\cdot, W) \right.$$
$$\left. + h(R(\cdot, W)X, \cdot) + h(R(\cdot, W)\cdot, X) \right]$$
$$+ \frac{1}{2} \mathrm{tr} \left[\nabla_{W,\cdot}^2 h(X, \cdot) - \nabla_{X,W}^2 h(\cdot, \cdot) \right.$$
$$\left. + \nabla_{X,\cdot}^2 h(\cdot, W) - \nabla_{\cdot,\cdot}^2 h(X, W) \right]$$
$$= -\frac{1}{2} \mathrm{tr} \left[h(R(X, \cdot)W, \cdot) + h(R(X, \cdot)\cdot, W) \right.$$
$$\left. + h(R(W, \cdot)X, \cdot) + h(R(W, \cdot)\cdot, X) \right]$$
$$- \frac{1}{2} \left[(\nabla \delta h)(X, W) + \mathrm{Hess}(\mathrm{tr} h)(X, W) \right.$$
$$\left. + (\nabla \delta h)(W, X) + (\Delta h)(X, W) \right].$$

The first and third terms in this final expression are equal, thanks to (2.3.12) and the symmetries of Rm and h. The second and fourth terms are handled with (2.3.11). Applying also (2.3.8), we get the simplified expression

$$\frac{\partial}{\partial t}\text{Ric}(X, W) = -\text{tr}\, h(R(X, \cdot)W, \cdot) + \frac{1}{2}\, [h(W, \text{Ric}(X)) + h(X, \text{Ric}(W))]$$

$$- \frac{1}{2}[\mathcal{L}_{(\delta h)^\#}g(X, W) + \text{Hess}(\text{tr}h)(X, W) + (\Delta h)(X, W)],$$

which together with (2.3.10) is what we wanted to prove. $\qquad\square$

Proof. (Proposition 2.3.9.) By Proposition 2.3.6, we have

$$\frac{\partial R}{\partial t} = \frac{\partial}{\partial t}(\text{trRic}) = -\langle h, \text{Ric}\rangle + \text{tr}\left(\frac{\partial}{\partial t}\text{Ric}\right).$$

From Proposition 2.3.7, and (2.3.10) we then see that

$$\frac{\partial R}{\partial t} = -\langle h, \text{Ric}\rangle - \frac{1}{2}\text{tr}\Delta_L - \frac{1}{2}\text{tr}\mathcal{L}_{(\delta h)^\#}g - \frac{1}{2}\text{tr}\,\text{Hess}(\text{tr}\, h). \qquad (2.3.15)$$

Expanding the definition of Δ_L using (2.3.11) and (2.3.12) gives

$$\Delta_L(X, W) = \Delta h(X, W) - \langle h(X, \cdot), \text{Ric}(W, \cdot)\rangle$$

$$- \langle h(W, \cdot), \text{Ric}(X, \cdot)\rangle + 2\langle \text{Rm}(X, \cdot, W, \cdot), h\rangle,$$

making it easier to see that

$$\text{tr}\Delta_L h = \text{tr}\Delta h - \langle h, \text{Ric}\rangle - \langle h, \text{Ric}\rangle + 2\langle h, \text{Ric}\rangle = \Delta(\text{tr}\, h). \qquad (2.3.16)$$

Meanwhile, by (2.3.8),

$$\text{tr}\mathcal{L}_{(\delta h)^\#}g = -2\delta^2 h. \qquad (2.3.17)$$

Combining (2.3.15), (2.3.16) and (2.3.17) gives our conclusion

$$\frac{\partial R}{\partial t} = -\langle h, \text{Ric}\rangle + \delta^2 h - \Delta(\text{tr}h).$$

$\qquad\square$

Proof. (Proposition 2.3.10.) The divergence theorem tells us that for a fixed 1-form α,

$$\int (\delta\alpha)\, dV = 0.$$

(Note that $\delta = (-1)^{n(p+1)+1} * d*$ on a p-form, and hence $(\delta\alpha)dV = (\pm * d * \alpha)dV = \pm d(*\alpha)$.) This enables us to integrate by parts: if $f : \mathcal{M} \to \mathbb{R}$, then

$$\delta(f\alpha) = -\langle df, \alpha\rangle + f(\delta\alpha) \qquad (2.3.18)$$

so by integrating,

$$\int \langle df, \alpha \rangle \, dV = \int f(\delta\alpha) \, dV. \tag{2.3.19}$$

Applying this formula with α equal to our time-dependent 1-form ω, and differentiating with respect to t, gives

$$\int \frac{\partial}{\partial t}(\delta\omega) f \, dV = - \int h(df, \omega) \, dV + \int \left\langle df, \frac{\partial\omega}{\partial t} \right\rangle dV$$
$$- \int [(\delta\omega) f - \langle df, \omega \rangle] \frac{1}{2}(\mathrm{tr} h) \, dV$$

where we have used $\frac{\partial}{\partial t} g^{ij} = -h^{ij}$ to obtain the sign in the first term on the right-hand side.

By applying (2.3.18) with $\alpha = \omega$ to the final term, and (2.3.19) with $\alpha = \frac{\partial\omega}{\partial t}$ to the penultimate term,

It follows that

$$0 = \int \frac{\partial}{\partial t}(\delta\omega) f \, dV + \int \langle df, h(\omega, \cdot) \rangle \, dV$$
$$- \int f \left(\delta \frac{\partial\omega}{\partial t} \right) dV + \int [\delta(f\omega)] \frac{1}{2}(\mathrm{tr} h) \, dV$$
$$= \int \frac{\partial}{\partial t}(\delta\omega) f \, dV + \int f\delta\big(h(\omega, \cdot)\big) \, dV$$
$$- \int f \left(\delta \frac{\partial\omega}{\partial t} \right) dV + \int \left\langle d\left(\frac{\mathrm{tr} h}{2}\right), \omega f \right\rangle dV$$
$$= \int \left[\frac{\partial}{\partial t}(\delta\omega) + \langle \delta h, \omega \rangle - \langle h, \nabla\omega \rangle - \delta\frac{\partial\omega}{\partial t} + \left\langle d\left(\frac{\mathrm{tr} h}{2}\right), \omega \right\rangle \right] f \, dV$$

for any f. Therefore

$$\frac{\partial}{\partial t}(\delta\omega) = \delta\frac{\partial\omega}{\partial t} + \langle h, \nabla\omega \rangle - \langle \delta h, \omega \rangle - \left\langle d\left(\frac{\mathrm{tr} h}{2}\right), \omega \right\rangle$$
$$= \delta\frac{\partial\omega}{\partial t} + \langle h, \nabla\omega \rangle - \langle \delta G(h), \omega \rangle .$$

$\square$

Proof. (Proposition 2.3.11.)   Let us first note that for applications of this proposition in these notes, we only need know that

$$\left(\frac{\partial}{\partial t} \delta G(T) \right) Z = -T((\delta G(h))^{\#}, Z) + \text{terms involving no derivatives of } h.$$

As usual, we need the formula $\mathcal{L}_Z g(X, Y) = \langle \nabla_X Z, Y \rangle + \langle X, \nabla_Y Z \rangle$, analogous to (2.3.8).

In particular, we have the analogue of (2.3.18) in the proof of the last proposition, that for all $S \in \Gamma(\mathrm{Sym}^2 T^*\mathcal{M})$

$$(\delta S)Z = \delta(S(\cdot, Z)) + \frac{1}{2}\langle S, \mathcal{L}_Z g \rangle, \qquad (2.3.20)$$

and also, by differentiating, and applying Proposition 2.3.1

$$\frac{\partial}{\partial t}\mathcal{L}_Z g(X, Y) = h(\nabla_X Z, Y) + h(X, \nabla_Y Z) + \nabla_Z h(X, Y). \qquad (2.3.21)$$

By the definition $G(T) = T - \frac{1}{2}(\mathrm{tr}T)g$ of (2.1.7) and Proposition 2.3.6 we have

$$\frac{\partial G(T)}{\partial t} = -\frac{1}{2}\frac{\partial \mathrm{tr}T}{\partial t}g - \frac{1}{2}(\mathrm{tr}T)h = \frac{1}{2}\langle h, T \rangle g - \frac{1}{2}(\mathrm{tr}T)h. \qquad (2.3.22)$$

Having compiled these preliminary formulae, we apply (2.3.20) with $S = G(T)$, and differentiate with respect to t to give

$$\left(\frac{\partial}{\partial t}\delta G(T)\right)Z = \frac{\partial}{\partial t}\delta\left(G(T)(\cdot, Z)\right) + \frac{1}{2}\frac{\partial}{\partial t}\langle G(T), \mathcal{L}_Z g \rangle. \qquad (2.3.23)$$

We deal with the two terms on the right-hand side separately. For the first, by Proposition 2.3.10 and (2.3.22),

$$\frac{\partial}{\partial t}\delta\left(G(T)(\cdot, Z)\right) = \delta\frac{\partial}{\partial t}\left(G(T)(\cdot, Z)\right) + \langle h, \nabla(G(T)(\cdot, Z)) \rangle$$
$$- \langle \delta G(h), G(T)(\cdot, Z) \rangle$$
$$= \delta\left[\frac{1}{2}(\langle h, T \rangle g - (\mathrm{tr}T)h)(\cdot, Z)\right] + \langle h, \nabla G(T)(\cdot, \cdot, Z) \rangle$$
$$+ \langle h, G(T)(\cdot, \nabla Z) \rangle - \langle \delta G(h), T(\cdot, Z) - \frac{1}{2}(\mathrm{tr}T)g(\cdot, Z) \rangle$$
$$= -\frac{1}{2}Z\langle h, T \rangle + \frac{1}{2}\langle h, T \rangle \delta(g(\cdot, Z)) + h\left(\nabla\left(\frac{\mathrm{tr}T}{2}\right), Z\right)$$
$$- \frac{1}{2}(\mathrm{tr}T)\delta(h(\cdot, Z)) + \langle h, \nabla T(\cdot, \cdot, Z) \rangle$$
$$+ \left\langle h, -\frac{1}{2}d(\mathrm{tr}T) \otimes g(\cdot, Z) \right\rangle$$
$$+ \langle h, G(T)(\cdot, \nabla Z) \rangle - T((\delta G(h))^{\#}, Z)$$
$$+ \frac{1}{2}(\mathrm{tr}T)(\delta G(h))Z.$$

Using (2.3.20) with $S = g$ and also with $S = h$, and recalling (2.1.8), we then have

$$\frac{\partial}{\partial t}\delta\left(G(T)(\cdot, Z)\right) = -\frac{1}{2}\langle\nabla_Z h, T\rangle - \frac{1}{2}\langle h, \nabla_Z T\rangle - \frac{1}{4}\langle h, T\rangle\mathrm{tr}\mathcal{L}_Z g$$

$$- \frac{1}{2}(\mathrm{tr}T)\left[(\delta h)Z - \frac{1}{2}\langle h, \mathcal{L}_Z g\rangle\right] + \langle h, \nabla T(\cdot, \cdot, Z)\rangle$$

$$+ \langle h, G(T)(\cdot, \nabla.Z)\rangle - T((\delta G(h))^{\#}, Z) + \frac{1}{2}(\mathrm{tr}T)(\delta h)Z$$

$$+ \frac{1}{4}(\mathrm{tr}T)Z(\mathrm{tr}h)$$

$$= \left(-T((\delta G(h))^{\#}, Z) + \langle h, \nabla T(\cdot, \cdot, Z)\rangle - \frac{1}{2}\langle h, \nabla_Z T\rangle\right)$$

$$- \frac{1}{2}\langle\nabla_Z h, G(T)\rangle - \frac{1}{4}\langle h, T\rangle\mathrm{tr}\mathcal{L}_Z g + \frac{1}{4}(\mathrm{tr}T)\langle h, \mathcal{L}_Z g\rangle$$

$$+ \langle h, G(T)(\cdot, \nabla.Z)\rangle.$$

For the second term of (2.3.23), we recall (2.3.22) and (2.3.21), and think in the same manner as in the proof of Proposition 2.3.6, to compute

$$\frac{1}{2}\frac{\partial}{\partial t}\langle G(T), \mathcal{L}_Z g\rangle = \frac{1}{2}\left\langle\frac{\partial}{\partial t}G(T), \mathcal{L}_Z g\right\rangle + \frac{1}{2}\left\langle G(T), \frac{\partial}{\partial t}\mathcal{L}_Z g\right\rangle$$

$$- \langle h, G(T)(\cdot, \nabla.Z)\rangle - \langle G(T), h(\cdot, \nabla.Z)\rangle$$

$$= \frac{1}{2}\left\langle\frac{1}{2}\langle h, T\rangle g - \frac{1}{2}(\mathrm{tr}T)h, \mathcal{L}_Z g\right\rangle$$

$$+ \frac{1}{2}\langle G(T), 2h(\nabla.Z, \cdot) + \nabla_Z h\rangle$$

$$- \langle h, G(T)(\cdot, \nabla.Z)\rangle - \langle G(T), h(\cdot, \nabla.Z)\rangle$$

$$= \frac{1}{4}\langle h, T\rangle\mathrm{tr}\mathcal{L}_Z g - \frac{1}{4}(\mathrm{tr}T)\langle h, \mathcal{L}_Z g\rangle$$

$$+ \frac{1}{2}\langle G(T), \nabla_Z h\rangle - \langle h, G(T)(\cdot, \nabla.Z)\rangle.$$

Combining these two formulae with (2.3.23), we conclude

$$\left(\frac{\partial}{\partial t}\delta G(T)\right)Z = -T((\delta G(h))^{\#}, Z) + \langle h, \nabla T(\cdot, \cdot, Z)\rangle - \frac{1}{2}\langle h, \nabla_Z T\rangle.$$

$\square$

Proof. (Proposition 2.3.12.)   This follows easily from the standard formula for the derivative of a t-dependent matrix $A(t)$,

$$\frac{d}{dt}\ln\det A(t) = \mathrm{tr}\left(A(t)^{-1}\frac{dA(t)}{dt}\right),$$

or by direct computation in normal coordinates.

$\square$

2.4 Laplacian of the curvature tensor

A reference for this section is [19, Lemma 7.2]. We define the tensor $B \in \Gamma(\otimes^4 T^* \mathcal{M})$ by

$$B(X, Y, W, Z) = \langle \mathrm{Rm}(X, \cdot, Y, \cdot), \mathrm{Rm}(W, \cdot, Z, \cdot) \rangle,$$

which has some but not all of the symmetries of the curvature tensor:

$$B(X, Y, W, Z) = B(W, Z, X, Y) = B(Y, X, Z, W).$$

Proposition 2.4.1.

$$
\begin{aligned}
(\Delta \mathrm{Rm})(X, Y, W, Z) = &-\nabla^2_{Y,W} \mathrm{Ric}(X, Z) + \nabla^2_{X,W} \mathrm{Ric}(Y, Z) \\
&- \nabla^2_{X,Z} \mathrm{Ric}(Y, W) + \nabla^2_{Y,Z} \mathrm{Ric}(X, W) \\
&- \mathrm{Ric}(R(W, Z)Y, X) + \mathrm{Ric}(R(W, Z)X, Y) \\
&- 2(B(X, Y, W, Z) - B(X, Y, Z, W) \\
&+ B(X, W, Y, Z) - B(X, Z, Y, W))
\end{aligned}
$$

Proof. This sort of calculation is probably easiest to perform in normal coordinates (or alternatively with respect to an appropriate orthonormal frame) about an arbitrary point p. The main ingredients are the Bianchi identities. To begin with, we require the second Bianchi identity for the first derivative of Rm,

$$\nabla_i R_{jkla} + \nabla_j R_{kila} + \nabla_k R_{ijla} = 0.$$

(Here $\nabla_i R_{jkla} := (\nabla_{\frac{\partial}{\partial x^i}} \mathrm{Rm})(\frac{\partial}{\partial x^j}, \frac{\partial}{\partial x^k}, \frac{\partial}{\partial x^l}, \frac{\partial}{\partial x^a})$.) Taking one further derivative, tracing, and restricting our attention to the point p, we find that

$$\Delta R_{jkla} + \nabla_i \nabla_j R_{kila} - \nabla_i \nabla_k R_{jila} = 0. \tag{2.4.1}$$

(Here $\Delta R_{jkla} := (\Delta \mathrm{Rm})(\frac{\partial}{\partial x^j}, \frac{\partial}{\partial x^k}, \frac{\partial}{\partial x^l}, \frac{\partial}{\partial x^a})$.) Note that when considering expressions at the point p, where the vectors $\{\frac{\partial}{\partial x^i}\}$ are orthonormal, we are able to use only lower indices, and the usual summation convention makes sense.

We focus on the second term of (2.4.1), since the third term differs only by a sign and a permutation of k and j. By the Ricci identity (2.1.2) we have

$$\nabla_i \nabla_j R_{kila} - \nabla_j \nabla_i R_{kila} = -R_{jikc} R_{cila} - R_{jiic} R_{kcla} - R_{jilc} R_{kica} - R_{jiac} R_{kilc}.$$

Now $R_{jiic} = -R_{jc}$ with our sign convention, and we have the first Bianchi identity $R_{cila} = -R_{ilca} - R_{lcia}$, and thus

$$
\begin{aligned}
\nabla_i \nabla_j R_{kila} - \nabla_j \nabla_i R_{kila} = &\; R_{jikc} R_{ilca} + R_{jikc} R_{lcia} + R_{jc} R_{kcla} \\
&- R_{jilc} R_{kica} - R_{jiac} R_{kilc} \\
= &\; R_{jc} R_{kcla} + B_{jkla} - B_{jkal} + B_{jlka} - B_{jakl}.
\end{aligned}
$$

$$\tag{2.4.2}$$

To handle the second term on the left-hand side of (2.4.2), we return to the second Bianchi identity, with permuted indices

$$\nabla_b R_{laki} + \nabla_l R_{abki} + \nabla_a R_{blki} = 0,$$

near p, and trace to give

$$g^{bi}\nabla_b R_{kila} + \nabla_l R_{ak} - \nabla_a R_{lk} = 0.$$

Applying ∇_j and restricting our attention to p, we see that

$$\nabla_j\nabla_i R_{kila} = \nabla_j\nabla_a R_{lk} - \nabla_j\nabla_l R_{ak},$$

which may be plugged into (2.4.2) to give

$$\nabla_i\nabla_j R_{kila} = \nabla_j\nabla_a R_{lk} - \nabla_j\nabla_l R_{ak} + R_{jc}R_{kcla}$$
$$+ B_{jkla} - B_{jkal} + B_{jlka} - B_{jakl}.$$

Finally, we apply this twice to (2.4.1) (once with k and j permuted) to conclude

$$\Delta R_{jkla} = -\nabla_j\nabla_a R_{lk} + \nabla_j\nabla_l R_{ak} + \nabla_k\nabla_a R_{lj} - \nabla_k\nabla_l R_{aj}$$
$$-R_{jc}R_{kcla} + R_{kc}R_{jcla} - 2(B_{jkla} - B_{jkal} + B_{jlka} - B_{jakl}).$$

$$\square$$

2.5 Evolution of curvature and geometric quantities under Ricci flow

In this section, we plug $h = -2\text{Ric}$ into the formulae of Section 2.3.1 describing how geometric quantities such as curvature evolve under arbitrary variations of the metric, and simplify the resulting expressions.

Proposition 2.3.5 in the case $h = -2\text{Ric}$ may be simplified using Proposition (2.4.1) to immediately give the following formula.

Proposition 2.5.1. *Under the Ricci flow, the curvature tensor evolves according to*

$$\frac{\partial}{\partial t}\text{Rm}(X, Y, W, Z) = (\Delta\text{Rm})(X, Y, W, Z)$$
$$- \text{Ric}(R(X, Y)W, Z) + \text{Ric}(R(X, Y)Z, W)$$
$$- \text{Ric}(R(W, Z)X, Y) + \text{Ric}(R(W, Z)Y, X) \quad (2.5.1)$$
$$+ 2(B(X, Y, W, Z) - B(X, Y, Z, W)$$
$$+ B(X, W, Y, Z) - B(X, Z, Y, W)).$$

Therefore, the curvature tensor Rm evolves under a heat equation.

Remark 2.5.2. A concise way of writing this, which contains enough information for the applications we have in mind, is:

$$\frac{\partial}{\partial t}\text{Rm} = \Delta \text{Rm} + \text{Rm} * \text{Rm} \qquad (2.5.2)$$

where we are using the $*$-notation from Section 2.1.

We can also compute pleasing expressions for the evolution of the Ricci and scalar curvatures. For the former, it is easiest to set $h = -2\,\text{Ric}$ in Proposition 2.3.7, rather than working directly from Proposition 2.5.1. Keeping in mind that $\delta G(\text{Ric}) = 0$ by (2.1.9), we find the following formula.

Proposition 2.5.3. *Under the Ricci flow, the Ricci tensor evolves according to*

$$\frac{\partial}{\partial t}\text{Ric} = \Delta_L(\text{Ric}), \qquad (2.5.3)$$

or equivalently,

$$\frac{\partial}{\partial t}\text{Ric}(X, W) = \Delta \text{Ric}(X, W) - 2\,\langle\text{Ric}(X), \text{Ric}(W)\rangle + 2\,\langle\text{Rm}(X, \cdot, W, \cdot), \text{Ric}\rangle.$$
$$(2.5.4)$$

Meanwhile, for the scalar curvature, it is easiest to work directly from Proposition 2.3.9. Setting $h = -2\,\text{Ric}$ again, and keeping in mind that by the contracted second Bianchi identity (2.1.9) we have $\delta^2 \text{Ric} = \frac{1}{2}\Delta R$, we find the following.

Proposition 2.5.4. *Under the Ricci flow, the scalar curvature evolves according to*

$$\frac{\partial R}{\partial t} = \Delta R + 2|\text{Ric}|^2. \qquad (2.5.5)$$

By making the orthogonal decomposition

$$\text{Ric} = \overset{\circ}{\text{Ric}} + \frac{R}{n}g$$

of the Ricci curvature in terms of the traceless Ricci curvature $\overset{\circ}{\text{Ric}}$, we see that

$$|\text{Ric}|^2 = \left|\overset{\circ}{\text{Ric}}\right|^2 + \frac{R^2}{n^2}|g|^2 \geq 0 + \frac{R^2}{n},$$

which gives us the following differential inequality for R:

Corollary 2.5.5.

$$\frac{\partial R}{\partial t} \geq \Delta R + \frac{2}{n}R^2. \qquad (2.5.6)$$

Let us also specialise Proposition 2.3.10 to the Ricci flow, and the situation where ω is exact. Using the contracted second Bianchi identity (2.1.9) again, and keeping in mind our sign convention $\Delta = -\delta d$, we immediately obtain:

Proposition 2.5.6. *If $f : \mathcal{M} \to \mathbb{R}$ is a time dependent function, then under the Ricci flow,*

$$\frac{\partial}{\partial t}\Delta f = \Delta \frac{\partial f}{\partial t} + 2\langle \mathrm{Ric}, \mathrm{Hess}(f) \rangle.$$

Finally, the evolution of the volume under Ricci flow follows immediately from Proposition 2.3.12:

$$\frac{\partial}{\partial t} dV = -R\, dV. \tag{2.5.7}$$

In particular, writing $V(t) := \mathrm{Vol}((\mathcal{M}, g(t)))$, we have

$$\frac{dV}{dt} = -\int R\, dV. \tag{2.5.8}$$

3

The maximum principle

3.1 Statement of the maximum principle

Theorem 3.1.1 (Weak maximum principle for scalars). *Suppose for $t \in$ $[0, T]$ (where $0 < T < \infty$) that $g(t)$ is a smooth family of metrics, and $X(t)$ is a smooth family of vector fields on a closed manifold $\mathcal{M}$. Let $F : \mathbb{R} \times [0, T] \to \mathbb{R}$ be smooth. Suppose that $u \in C^\infty(\mathcal{M} \times [0, T], \mathbb{R})$ solves*

$$\frac{\partial u}{\partial t} \leq \Delta_{g(t)} u + \langle X(t), \nabla u \rangle + F(u, t). \tag{3.1.1}$$

Suppose further that $\phi : [0, T] \to \mathbb{R}$ solves

$$\begin{cases} \frac{d\phi}{dt} &= F(\phi(t), t) \\ \phi(0) &= \alpha \in \mathbb{R}. \end{cases} \tag{3.1.2}$$

If $u(\cdot, 0) \leq \alpha$, then $u(\cdot, t) \leq \phi(t)$ for all $t \in [0, T]$.

By applying this result with the signs of u, ϕ and α reversed, and F appropriately modified, we find the following modification:

Corollary 3.1.2 (Weak minimum principle). *Theorem 3.1.1 also holds with the sense of all three inequalities reversed (that is, replacing all three instances of $\leq$ by $\geq$).*

Proof. (Theorem 3.1.1.) For $\varepsilon > 0$, consider the ODE

$$\begin{cases} \frac{d\phi_\varepsilon}{dt} &= F(\phi_\varepsilon(t), t) + \varepsilon \\ \phi_\varepsilon(0) &= \alpha + \varepsilon \in \mathbb{R}, \end{cases} \tag{3.1.3}$$

for a new function $\phi_\varepsilon : [0, T] \to \mathbb{R}$. Basic ODE theory tells us that there exists $\varepsilon_0 > 0$ such that for $0 < \varepsilon \leq \varepsilon_0$ there exists a solution ϕ_ε on $[0, T]$. (Here we are using the existence of ϕ asserted in the hypotheses, and the fact that $T < \infty$.)

Moreover, $\phi_\varepsilon \to \phi$ uniformly as $\varepsilon \downarrow 0$. Consequently, we need only prove that $u(\cdot, t) < \phi_\varepsilon(t)$ for all $t \in [0, T]$ and arbitrary $\varepsilon \in (0, \varepsilon_0)$.

If this were not true, then we could choose $\varepsilon \in (0, \varepsilon_0)$ and $t_0 \in (0, T]$ where $u(\cdot, t_0) < \phi_\varepsilon(t_0)$ fails. Without loss of generality we may assume that t_0 is the earliest such time, and pick $x \in \mathcal{M}$ so that $u(x, t_0) = \phi_\varepsilon(t_0)$. Using the fact that $u(x, s) - \phi_\varepsilon(s)$ is negative for $s \in [0, t_0)$ and zero for $s = t_0$, we must have

$$\frac{\partial u}{\partial t}(x, t_0) - \phi'_\varepsilon(t_0) \geq 0.$$

Moreover, since x is a maximum of $u(\cdot, t_0)$, it follows that $\Delta u(x, t_0) \leq 0$ and $\nabla u(x, t_0) = 0$.

Combining these facts with the inequality (3.1.1) for u and equation (3.1.3) for ϕ_ε, we get the contradiction that

$$0 \geq \left[\frac{\partial u}{\partial t} - \Delta u - \langle X, \nabla u \rangle - F(u, \cdot) \right](x, t_0) \geq \phi'_\varepsilon(t_0) - F(\phi_\varepsilon(t_0), t_0) = \varepsilon > 0.$$

$\square$

Remark 3.1.3.   The *strong* maximum principle for scalars tells us that in fact $u(\cdot, t) < \phi(t)$ for all $t \in (0, T]$, *unless* $u(x, t) = \phi(t)$ for all $x \in \mathcal{M}$ and $t \in [0, T]$.

3.2 Basic control on the evolution of curvature

By applying the maximum principle to various equations and inequalities governing the evolution of curvature, we will get some preliminary control on how R and Rm evolve. Chapter 9 will be dedicated to obtaining more refined estimates.

Theorem 3.2.1.   *Suppose $g(t)$ is a Ricci flow on a closed manifold $\mathcal{M}$, for $t \in [0, T]$. If $R \geq \alpha \in \mathbb{R}$ at time $t = 0$, then for all times $t \in [0, T]$,*

$$R \geq \frac{\alpha}{1 - \left(\frac{2\alpha}{n}\right)t} \tag{3.2.1}$$

Proof.   Simply apply the weak minimum principle (Corollary 3.1.2) to (2.5.6) with $u \equiv R$, $X \equiv 0$ and $F(r, t) \equiv \frac{2}{n}r^2$. In this case,

$$\phi(t) = \frac{\alpha}{1 - \left(\frac{2\alpha}{n}\right)t}.$$

$\square$

Corollary 3.2.2. *Suppose $g(t)$ is a Ricci flow on a closed manifold $\mathcal{M}$, for $t \in [0, T]$. If $R \geq \alpha \in \mathbb{R}$ at time $t = 0$, then $R \geq \alpha$ for all times $t \in [0, T]$.*

Corollary 3.2.3. *Positive (or weakly positive) scalar curvature is preserved under such a Ricci flow.*

Corollary 3.2.4. *Suppose $g(t)$ is a Ricci flow on a closed manifold $\mathcal{M}$, for $t \in [0, T)$. If $R \geq \alpha > 0$ at time $t = 0$, then we must have $T \leq \frac{n}{2\alpha}$.*

Corollary 3.2.5. *Suppose $g(t)$ is a Ricci flow on a closed manifold $\mathcal{M}$, for $t \in (0, T]$. Then*

$$R \geq -\frac{n}{2t},$$

for all $t \in (0, T]$.

By recalling the formula (2.5.8) for the evolution of the volume, we find from Corollary 3.2.3 that:

Corollary 3.2.6. *If $g(t)$ is a Ricci flow on a closed manifold $\mathcal{M}$, for $t \in [0, T]$, with $R \geq 0$ at $t = 0$, then the volume $V(t)$ is (weakly) decreasing.*

Corollary 3.2.7. *Suppose $g(t)$ is a Ricci flow on a closed manifold $\mathcal{M}$, for $t \in [0, T]$. If at time $t = 0$, we have $\alpha := \inf R < 0$ then*

$$\frac{V(t)}{\left(1 + \frac{2(-\alpha)}{n}t\right)^{\frac{n}{2}}} \tag{3.2.2}$$

is weakly decreasing and in particular

$$V(t) \leq V(0) \left(1 + \frac{2(-\alpha)}{n}t\right)^{\frac{n}{2}}. \tag{3.2.3}$$

Proof. We calculate

$$\frac{d}{dt} \ln \left[\frac{V(t)}{\left(1 + \frac{2(-\alpha)}{n}t\right)^{\frac{n}{2}}} \right] = \frac{d}{dt} \left[\ln V - \frac{n}{2} \ln \left(1 + \frac{2(-\alpha)}{n}t\right) \right]$$

$$= \frac{1}{V}\frac{dV}{dt} - \frac{-\alpha}{1 + \frac{2(-\alpha)}{n}t}$$

$$= -\frac{1}{V} \int R \, dV + \frac{\alpha}{1 + \frac{2(-\alpha)}{n}t}$$

$$\leq -\inf_{\mathcal{M}} R(t) + \frac{\alpha}{1 - \frac{2\alpha}{n}t} \leq 0$$

by Theorem 3.2.1. $\square$

Remark 3.2.8. One consequence of Corollary 3.2.7 is the following: If our Ricci flow is defined for all $t \in [0, \infty)$, then

$$\overline{V} := \lim_{t \to \infty} \frac{V(t)}{\left(1 + \frac{2(-\alpha)}{n} t\right)^{\frac{n}{2}}}$$

exists. A rough principle within the work of Perelman is a local version of an assertion that if $\overline{V} > 0$ then the manifold is becoming hyperbolic. (If $\overline{V} = 0$ then the manifold is becoming like a graph manifold.) Recall from the discussion in Section 1.2.1 that if g_0 is exactly hyperbolic then $g(t) = (1 + 2(n-1)t)g_0$, so $V(t) = (1 + 2(n-1)t)^{\frac{n}{2}} V(0)$. See [32] for more information.

Remark 3.2.9. The scalar curvature is just one of many types of curvature for which positivity is preserved under the Ricci flow on closed manifolds. Other notable preserved curvature conditions include positive curvature operator in any dimension [20], positive isotropic curvature in four dimensions [23], and as we shall see in Chapter 9, in dimension three (or less) positive Ricci curvature. In contrast, negative curvatures are not preserved in general under Ricci flow (with the notable exception of scalar curvature in dimension two).

The maximum principle can also be applied to give constraints on the full curvature tensor Rm. For now, we concentrate on the *norm* of this tensor, which is comparable to the norm of the largest sectional curvature K (that is, $\frac{1}{C}|K| \leq |\text{Rm}| \leq C|K|$ for some constant C depending only on n). Later, in Chapter 9, we will obtain more refined control.

Proposition 3.2.10. *Under Ricci flow*

$$\frac{\partial}{\partial t}|\text{Rm}|^2 \leq \Delta|\text{Rm}|^2 - 2|\nabla\text{Rm}|^2 + C|\text{Rm}|^3 \qquad (3.2.4)$$

where $C = C(n)$ and Δ denotes the Laplace-Beltrami operator.

Proof. Since $\frac{\partial}{\partial t} g^{ij} = -h^{ij} = 2R^{ij}$, we may apply Remark 2.5.2 and calculate

$$\frac{\partial}{\partial t}|\text{Rm}|^2 = \frac{\partial}{\partial t}\left(g^{ij}g^{kl}g^{ab}g^{cd}R_{ikac}R_{jlbd}\right)$$

$$= 2\left(R^{ij}g^{kl}g^{ab}g^{cd} + g^{ij}R^{kl}g^{ab}g^{cd} + g^{ij}g^{kl}R^{ab}g^{cd}\right.$$

$$\left. + g^{ij}g^{kl}g^{ab}R^{cd}\right)R_{ikac}R_{jlbd} + 2\left\langle\text{Rm}, \frac{\partial}{\partial t}\text{Rm}\right\rangle$$

$$\leq C|\text{Rm}|^3 + 2\langle\text{Rm}, \Delta\text{Rm} + \text{Rm} * \text{Rm}\rangle$$

$$\leq 2\langle\text{Rm}, \Delta\text{Rm}\rangle + C|\text{Rm}|^3.$$

However $d|\text{Rm}|^2 = 2\langle\text{Rm}, \nabla\text{Rm}\rangle$ so

$$\Delta|\text{Rm}|^2 = 2|\nabla\text{Rm}|^2 + 2\langle\text{Rm}, \Delta\text{Rm}\rangle.$$

$\square$

After weakening (3.2.4) to

$$\frac{\partial}{\partial t}|\text{Rm}|^2 \le \Delta|\text{Rm}|^2 + C|\text{Rm}|^3,$$

we may apply the weak maximum principle, Theorem 3.1.1, with $u = |\text{Rm}|^2$, $X \equiv 0$, $F(r, t) = C r^{\frac{3}{2}}$, $\alpha = M^2$ and

$$\phi(t) = \frac{1}{\left(M^{-1} - \frac{1}{2}Ct\right)^2},$$

a solution of

$$\begin{cases} \phi'(t) = C\phi(t)^{\frac{3}{2}} \\ \phi(0) = M^2, \end{cases}$$

to give:

Theorem 3.2.11. *Suppose $g(t)$ is a Ricci flow on a closed manifold $\mathcal{M}$, for $t \in [0, T]$, and that at $t = 0$ we have $|\text{Rm}| \le M$. Then for all $t \in (0, T]$,*

$$|\text{Rm}| \le \frac{M}{1 - \frac{1}{2}CMt},$$

where C is the same constant as in Proposition 3.2.10.

Remark 3.2.12. One interpretation of this theorem is that if $|\text{Rm}| \le 1$ at time $t = 0$ (as one can achieve by a rescaling as in Section 1.2.3) then the amount of time which must elapse before the maximum of $|\text{Rm}|$ doubles is bounded below by a positive constant dependent only on the dimension n.

In contrast, the scalar curvature can behave quite badly. We could have zero scalar curvature initially but not for $t > 0$, and by rescaling (again, as in Section 1.2.3) we may make the scalar curvature blow up as quickly as we like. An example would be the four dimensional flow starting at the product of a standard 2-sphere and a hyperbolic surface.

3.3 Global curvature derivative estimates

We have seen in Proposition 2.5.1 that the curvature tensor satisfies a heat equation. One property of such equations, as for elliptic equations, is that one can hope to control higher derivatives of the solution in terms of lower derivatives. In this section, we prove such derivative estimates over the whole manifold.

Theorem 3.3.1. *Suppose that $M > 0$ and $g(t)$ is a Ricci flow on a closed manifold $\mathcal{M}^n$ for $t \in [0, \frac{1}{M}]$. For all $k \in \mathbb{N}$ there exists $C = C(n, k)$ such that* *if* $|\text{Rm}| \le M$ *throughout* $\mathcal{M} \times [0, \frac{1}{M}]$, *then for all* $t \in [0, \frac{1}{M}]$,

$$|\nabla^k \text{Rm}| \le \frac{CM}{t^{\frac{k}{2}}}. \tag{3.3.1}$$

Proof.   We will just give a proof of the case $k = 1$ here. The higher derivative estimates follow along the same lines by induction.

By Remark 2.5.2 we have

$$\frac{\partial}{\partial t}\text{Rm} = \Delta\text{Rm} + \text{Rm} * \text{Rm}. \tag{3.3.2}$$

Specialising some of our formulae for commuting derivatives to our situation, we find from (2.1.6) that

$$\nabla(\Delta\text{Rm}) = \Delta(\nabla\text{Rm}) + \text{Rm} * \nabla\text{Rm},$$

and from (2.3.3) that

$$\nabla\frac{\partial}{\partial t}\text{Rm} = \frac{\partial}{\partial t}\nabla\text{Rm} + \text{Rm} * \nabla\text{Rm}.$$

Therefore, taking the covariant derivative of (3.3.2), we have

$$\frac{\partial}{\partial t}\nabla\text{Rm} = \Delta(\nabla\text{Rm}) + \text{Rm} * \nabla\text{Rm}. \tag{3.3.3}$$

Computing as in the proof of Proposition 3.2.10, we find that

$$\frac{\partial}{\partial t}|\nabla\text{Rm}|^2 \le \Delta|\nabla\text{Rm}|^2 - 2|\nabla^2\text{Rm}|^2 + C|\text{Rm}||\nabla\text{Rm}|^2 \\ \le \Delta|\nabla\text{Rm}|^2 + C|\text{Rm}||\nabla\text{Rm}|^2. \tag{3.3.4}$$

Set $u(x, t) = t|\nabla\text{Rm}|^2 + \alpha|\text{Rm}|^2$ for some α to be picked later. Then using (3.3.4) and Proposition 3.2.10, we have that

$$\frac{\partial u}{\partial t} \le |\nabla\text{Rm}|^2 + t\left(\Delta|\nabla\text{Rm}|^2 + C|\text{Rm}||\nabla\text{Rm}|^2\right) \\ + \alpha\left(\Delta|\text{Rm}|^2 - 2|\nabla\text{Rm}|^2 + C|\text{Rm}|^3\right) \\ = \Delta u + |\nabla\text{Rm}|^2(1 + Ct|\text{Rm}| - 2\alpha) + C\alpha|\text{Rm}|^3.$$

By hypothesis, $|\text{Rm}| \le M$ and $t \le \frac{1}{M}$. Thus

$$\frac{\partial u}{\partial t} \le \Delta u + |\nabla\text{Rm}|^2(1 + C - 2\alpha) + C\alpha M^3$$

for $C = C(n)$. For sufficiently large α (say $\alpha = \frac{1}{2}(1 + C)$) we have that

$$\frac{\partial u}{\partial t} \le \Delta u + C(n)M^3.$$

Note further that $u(\cdot, 0) = \alpha |\text{Rm}|^2 \leq \alpha M^2$. Therefore by the weak maximum principle, Theorem 3.1.1 (comparing with $\phi(t) = \alpha M^2 + Ct M^3$ satisfying $\phi'(t) = CM^3$ and $\phi(0) = \alpha M^2$) it follows that

$$u(\cdot, t) \leq \alpha M^2 + Ct M^3 \leq CM^2.$$

Hence

$$t |\nabla \text{Rm}|^2 \leq u(\cdot, t) \leq CM^2,$$

so

$$|\nabla \text{Rm}| \leq \frac{CM}{\sqrt{t}}$$

for some $C = C(n)$. $\qquad\square$

Corollary 3.3.2. *Suppose that $M > 0$ and $g(t)$ is a Ricci flow on a closed manifold $\mathcal{M}^n$ for $t \in [0, \frac{1}{M}]$. For all $j, k \in \{0\} \cup \mathbb{N}$, there exists a constant $C = C(j, k, n)$ such that if $|\text{Rm}| \leq M$ throughout $\mathcal{M} \times [0, \frac{1}{M}]$, then*

$$\left| \frac{\partial^j}{\partial t^j} \nabla^k \text{Rm} \right| \leq \frac{CM}{t^{j + \frac{k}{2}}}. \tag{3.3.5}$$

Proof. By rescaling the entire flow, as in Section 1.2.3, we first note that it suffices to prove the corollary at $t = 1$. Since this time lies within the interval $(0, \frac{1}{M}]$, we know that

$$M \leq 1. \tag{3.3.6}$$

By (2.5.2), we know that

$$\frac{\partial}{\partial t} \text{Rm} = \Delta \text{Rm} + \text{Rm} * \text{Rm}$$

and hence, since $M \leq 1$, we find at $t = 1$ that

$$\left| \frac{\partial}{\partial t} \text{Rm} \right| \leq |\Delta \text{Rm}| + |\text{Rm} * \text{Rm}| \leq C(M + M^2) \leq CM,$$

by Theorem 3.3.1. Meanwhile, by (3.3.3), we have

$$\frac{\partial}{\partial t} \nabla \text{Rm} = \Delta(\nabla \text{Rm}) + \text{Rm} * \nabla \text{Rm},$$

and may again use the fact that $M \leq 1$ to give, at $t = 1$,

$$\left| \frac{\partial}{\partial t} \nabla \text{Rm} \right| \leq C|\nabla^3 \text{Rm}| + C|\text{Rm}||\nabla \text{Rm}| \leq C(M \mid M^2) \leq CM.$$

We have dealt with the cases $(j, k) = (1, 0)$ and $(1, 1)$. To handle the case $j = 1$ and $k > 1$, we can use induction to show that $\frac{\partial}{\partial t} \nabla^k \text{Rm}$ can be written as a $*$-composition of terms involving Rm and its covariant derivatives, since by (2.3.3) we know that

$$\frac{\partial}{\partial t}\nabla^k \text{Rm} = \nabla \frac{\partial}{\partial t}\nabla^{k-1}\text{Rm} + \nabla^{k-1}\text{Rm} * \nabla\text{Rm}.$$

By taking norms, as for the first two cases, each of these terms is bounded at $t = 1$ as required.

We have dealt with the case $j = 1$, using the case $j = 0$ handled by Theorem 3.3.1. One can see with hindsight that this is the first inductive step to proving the corollary for all j. Indeed, by what we have seen above, any expression $\frac{\partial^{l+1}}{\partial t^{l+1}}\nabla^k\text{Rm}$ can immediately be written in terms of the lth t derivative of terms involving only Rm and its spacial covariant derivatives. □

Although we don't need them in these notes, we remark that with more effort, one can prove a *local* version of these estimates, assuming only local curvature bounds. (See [36].) In applications of the following result, U would typically be an open ball within some larger manifold.

Theorem 3.3.3 (Shi). *Suppose that $g(t)$ is a Ricci flow defined on an arbitrary (boundaryless) manifold U, for $t \in [0, T]$, with no requirement that $g(t)$ is complete. Suppose further that $|\text{Rm}| \leq M$ on $U \times [0, T]$, and that $p \in U$ and $r > 0$ are such that $\overline{B_{g(0)}(p, r)} \subset U$. Then*

$$|\nabla\text{Rm}(p, T)|^2 \leq C(n)M^2 \left(\frac{1}{r^2} + \frac{1}{T} + M \right).$$

The notation $B_{g(0)}(p, r)$ denotes a geodesic ball with centre p and radius r, with respect to the metric at $t = 0$.

Similar estimates hold for higher derivatives.

4
Comments on existence theory
for parabolic PDE

In the next chapter, we will try to justify the existence of a Ricci flow, over some short time interval, starting with a given smooth initial metric, on a closed manifold $\mathcal{M}$. We do this by reducing the problem to the solution of a parabolic equation.

In this chapter we will describe how to recognise a parabolic equation, and sketch the type of existence and uniqueness results one can prove for reasonable PDE of this type. This is not meant as a substitute for a serious PDE course, but rather as motivation for such a course.

4.1 Linear scalar PDE

Consider a second order PDE on $\Omega \subset \mathbb{R}^n$ for a function $u : \Omega \to \mathbb{R}$ of the form

$$\frac{\partial u}{\partial t} = a_{ij}\partial_i\partial_j u + b_i\partial_i u + cu \qquad (4.1.1)$$

where $\partial_i := \frac{\partial}{\partial x^i}$ and $a_{ij}, b_i, c : \Omega \to \mathbb{R}$ are smooth coefficients. Such an equation is called *parabolic* if a_{ij} is uniformly positive definite – that is, if there exists $\lambda > 0$ such that

$$a_{ij}\xi_i\xi_j \geq \lambda|\xi|^2 \qquad (4.1.2)$$

for all $\xi \in \mathbb{R}^n$. For example, the heat equation $\frac{\partial u}{\partial t} = \Delta u$ is a parabolic PDE.

This notion extends to manifolds. Let $\mathcal{M}$ be closed and consider the PDE for $u : \mathcal{M} \to \mathbb{R}$ given by

$$\frac{\partial u}{\partial t} = L(u) \qquad (4.1.3)$$

where $L : C^\infty(\mathcal{M}) \to C^\infty(\mathcal{M})$ can be written with respect to local coordinates $\{x^i\}$ as

$$L(u) = a_{ij}\partial_i\partial_j u + b_i\partial_i u + cu$$

43

for a_{ij}, b_i and c locally defined smooth, real coefficients. Equation (4.1.3) is said to be *parabolic* if a_{ij} is positive definite for all $x \in \mathcal{M}$. Note that this is well defined, independent of the choice of coordinates. (See also Section 4.2.)

There is a good theory for such equations. For example, given a smooth initial function $u_0 : \mathcal{M} \to \mathbb{R}$, there exists a smooth solution $u : \mathcal{M} \times [0, \infty) \to \mathbb{R}$ to

$$\begin{cases} \frac{\partial u}{\partial t} = L(u) & \text{on } \mathcal{M} \times [0, \infty) \\ u(0) = u_0 & \text{on } \mathcal{M}. \end{cases}$$

We also have uniqueness: Suppose that $\frac{\partial u}{\partial t} = L(u)$ and $\frac{\partial v}{\partial t} = L(v)$ on $\mathcal{M} \times [0, T]$. If either $u(0) = v(0)$ or $u(T) = v(T)$ holds, then $u(t) = v(t)$ for all $t \in [0, T]$.

4.2 The principal symbol

Define the *principal symbol* $\sigma(L) : T^*\mathcal{M} \to \mathbb{R}$ by

$$\sigma(L)(x, \xi) = a_{ij}(x)\xi_i\xi_j. \tag{4.2.1}$$

This is well defined, independent of the choice of coordinates. An alternative definition would be the following: Given $(x, \xi) \in T^*\mathcal{M}$, and $\phi, f : \mathcal{M} \to \mathbb{R}$ smooth with $d\phi(x) = \xi$, define

$$\sigma(L)(x, \xi)f(x) = \lim_{s \to \infty} s^{-2} e^{-s\phi(x)} L(e^{s\phi} f)(x). \tag{4.2.2}$$

To check that this is well defined, and that the two definitions are consistent, we calculate

$$\partial_i(e^{s\phi} f)(x) = s\partial_i\phi(x)(e^{s\phi} f)(x) + e^{s\phi}\partial_i f(x),$$

and so

$$e^{-s\phi}\partial_i\partial_j(e^{s\phi} f)(x) = s^2\partial_i\phi(x)\partial_j\phi(x)f(x) + s(\cdots) + s^0(\cdots)$$

where we have collected up terms according to whether they are quadratic, linear or constant in s, and we care only about the quadratic term. Therefore

$$e^{-s\phi(x)} L(e^{s\phi} f)(x) = s^2 a_{ij}\partial_i\phi(x)\partial_j\phi(x)f(x) + s(\cdots) + s^0(\cdots)$$

and so

$$\lim_{s \to \infty} s^{-2} e^{-s\phi(x)} L(e^{s\phi} f)(x) = a_{ij}\partial_i\phi(x)\partial_j\phi(x)f(x) = a_{ij}\xi_i\xi_j f(x).$$

Notice in particular that the second definition is independent of the choice of ϕ and f (with $d\phi(x) = \xi$).

Remark 4.2.1. Some authors would have $i\xi$ where we have ξ (c.f. the Fourier transform).

Thus the PDE (4.1.3) is parabolic if $\sigma(L)(x, \xi) > 0$ for all $(x, \xi) \in T^*\mathcal{M}$ with $\xi \neq 0$.

Example 4.2.2. Consider the Laplace Beltrami operator[1]

$$\Delta = \frac{1}{\sqrt{g}} \partial_i \left(\sqrt{g} g^{ij} \partial_j \right)$$

$$= g^{ij} \partial_i \partial_j + \text{lower order terms}$$

where $g := \det g_{ij}$. By definition,

$$\sigma(\Delta)(x, \xi) = g^{ij} \xi_i \xi_j = |\xi|^2 > 0.$$

and hence the heat equation $\frac{\partial u}{\partial t} = \Delta u$ is parabolic in this setting also.

4.3 Generalisation to Vector Bundles

The theory of parabolic equations extends under various generalisations of the concept of parabolic. For example, one can generalise to:

(i) higher order equations;
(ii) the case of less regular coefficients a_{ij} etc. and less regular u_0;

and more importantly for us, to:

(iii) time-dependent a_{ij};
(iv) u a section of a vector bundle;
(v) nonlinear equations.

We will now take a look at case (iv).

Let E be a smooth vector bundle over a closed manifold $\mathcal{M}$. Whereas above we looked at functions $u : \mathcal{M} \to \mathbb{R}$, we now consider sections $v \in \Gamma(E)$. Locally we may write $v = v^\alpha e_\alpha$ for some local frame $\{e_\alpha\}$. Consider the equation

$$\frac{\partial v}{\partial t} = L(v) \tag{4.3.1}$$

where L is a linear second order differential operator, by which we mean $L : \Gamma(E) \to \Gamma(E)$ may be given locally, in terms of local coordinates $\{x^i\}$ on $\mathcal{M}$

---

[1] In this chapter, only for this example do we attach the usual significance to upper and lower indices.

and a local frame $\{e_\alpha\}$ on E, as[2]

$$L(v) = \left[a_{\alpha\beta}^{ij}\partial_i\partial_j v^\beta + b_{\alpha\beta}^i\partial_i v^\beta + c_{\alpha\beta}v^\beta\right]e_\alpha.$$

We can again define the principal symbol – now $\sigma(L) : \Pi^*(E) \to \Pi^*(E)$ is a vector bundle homomorphism, where $\Pi : T^*\mathcal{M} \to \mathcal{M}$ denotes the bundle projection. Note that $\Pi^*(E)$ is a vector bundle over $T^*\mathcal{M}$ whose fibre at $(x, \xi) \in T^*(\mathcal{M})$ is E_x. We define

$$\sigma(L)(x, \xi)v = \left(a_{\alpha\beta}^{ij}\xi_i\xi_j v^\beta\right)e_\alpha. \tag{4.3.2}$$

Again, there exists a definition independent of coordinates: Given $(x, \xi) \in T^*\mathcal{M}$, for all $v \in \Gamma(E)$ and $\phi : \mathcal{M} \to \mathbb{R}$ with $d\phi(x) = \xi$, define

$$\sigma(L)(x, \xi)v = \lim_{s\to\infty} s^{-2}e^{-s\phi(x)}L(e^{s\phi}v)(x). \tag{4.3.3}$$

For example, for a vector bundle with connection ∇, if Δ denotes the connection Laplacian, then

$$\Delta(e^{s\phi}v) = e^{s\phi}(s^2|d\phi|^2 v + \text{lower order terms in } s),$$

and so $\sigma(\Delta)(x, \sigma)v = |\xi|^2 v$ because $d\phi = \xi$. Thus $\sigma(\Delta)(x, \xi) = |\xi|^2$ id.

We say that (4.3.1) is (strictly) parabolic if there exists $\lambda > 0$ such that

$$\langle\sigma(L)(x, \xi)v, v\rangle \geq \lambda|\xi|^2|v|^2 \tag{4.3.4}$$

for all $(x, \xi) \in T^*\mathcal{M}$ and $v \in \Gamma(E)$. This notion is also often called *strongly parabolic* to distinguish it from more general definitions.

Remark 4.3.1.   Equation (4.3.1) is parabolic if (4.3.4) is true for *any* fibre metric $\langle\cdot, \cdot\rangle$ on E.

Remark 4.3.2 (Special case).   Equation (4.3.1) is parabolic on the closed manifold $\mathcal{M}$ if for all $(x, \xi) \in T^*\mathcal{M}$ with $\xi \neq 0$, we have that $\sigma(L)(x, \xi)$ is a positive multiple of the identity on E_x.

In practice we need to be able to discuss *nonlinear* PDE of the form

$$\frac{\partial v}{\partial t} = P(v) \tag{4.3.5}$$

where P is a quasilinear second order differential operator, by which we mean $P : \Gamma(E) \to \Gamma(E)$ may be given locally, in terms of local coordinates $\{x^i\}$ on $\mathcal{M}$ and a local frame $\{e_\alpha\}$ on E, as

$$P(v) = \left[a_{\alpha\beta}^{ij}(x, v, \nabla v)\partial_i\partial_j v^\beta + b^\alpha(x, v, \nabla v)\right]e_\alpha.$$

---

[2] Don't try to read the usual significance into whether we use lower or upper indices.

Finally, the Ricci flow equation $\frac{\partial g}{\partial t} = -2\operatorname{Ric}(g)$ fits into this category, as we discuss in Section 5.1.

Given an arbitrary fixed section $w \in \Gamma(E)$ (independent of t) we call (4.3.5) *parabolic* (again, strictly and strongly) at w if the *linearisation* of (4.3.5) at w

$$\frac{\partial v}{\partial t} = [DP(w)]\, v$$

is parabolic as just defined.

4.4 Properties of parabolic equations

If $\frac{\partial v}{\partial t} = P(v)$ is parabolic at w (and the $a^{ij}_{\alpha\beta}$ and lower order terms satisfy some regularity hypotheses) then there exist $\varepsilon > 0$ and a smooth family $v(t) \in \Gamma(E)$ for $t \in [0, \varepsilon]$, such that

$$\begin{cases} \frac{\partial v}{\partial t} = P(v) & t \in [0, \varepsilon] \\ v(0) = w. \end{cases}$$

We also have uniqueness: Suppose that $\frac{\partial v}{\partial t} = P(v)$, $\frac{\partial w}{\partial t} = P(w)$ for $t \in [0, \varepsilon]$. If either $v(0) = w(0)$ or $v(\varepsilon) = w(\varepsilon)$, then $v(t) = w(t)$ for all $t \in [0, \varepsilon]$.

5

Existence theory for the Ricci flow

5.1 Ricci flow is not parabolic

Consider

$$\frac{\partial g}{\partial t} = Q(g) := -2\operatorname{Ric}(g), \tag{5.1.1}$$

on the bundle $E = \operatorname{Sym}^2 T^*\mathcal{M}$. The linearisation of (5.1.1) is[1]

$$\frac{\partial h}{\partial t} = Lh := [DQ(g)]\,h = \Delta_L h + \mathcal{L}_{(\delta G(h))^\sharp} g$$

by Proposition 2.3.7. We wish to know, for arbitrary initial metric g, and corresponding L, whether $\frac{\partial h}{\partial t} = L(h)$ is parabolic.

Recall that $\sigma(L)(x, \xi)h := \lim_{s \to \infty} s^{-2} e^{-s\phi(x)} L(e^{s\phi}h)(x)$ for ϕ satisfying $d\phi(x) = \xi$. Since $\Delta_L = \Delta + \text{lower order terms}$,

$$\sigma(\Delta_L)(x, \xi)h = \sigma(\Delta)(x, \xi)h = |\xi|^2 h$$

as calculated earlier. Meanwhile, we saw in (2.1.8) that $\delta G(h) = \delta h + \frac{1}{2}d(\operatorname{tr}h)$, and so

$$
\begin{aligned}
\delta G(e^{s\phi}h) &= -\operatorname{tr}\nabla(e^{s\phi}h) + \frac{1}{2}d(e^{s\phi}\operatorname{tr}h) \\
&= e^{s\phi}\left[s\left(-\operatorname{tr} d\phi \otimes h + \frac{1}{2}d\phi\operatorname{tr}h \right) + s^0(\cdots) \right] \\
&= e^{s\phi}\left[s\left(-h(\xi^\sharp, \cdot) + \frac{1}{2}\xi\operatorname{tr}h \right) + s^0(\cdots) \right]
\end{aligned}
$$

[1] Note that the subscript L for Δ denotes the Lichnerowicz Laplacian, and does not refer to the linearisation of Q.

48

Therefore, since $\mathcal{L}_{\omega^\sharp} g(X, W) = \nabla \omega(X, W) + \nabla \omega(W, X)$ as we saw in (2.3.8), we have

$$\mathcal{L}_{(\delta G(e^{s\phi}h))^\sharp} g = e^{s\phi}\big[s^2(-\xi \otimes h(\xi^\sharp, \cdot) - h(\xi^\sharp, \cdot) \otimes \xi + (\xi \otimes \xi)\mathrm{tr}h) \\ + s^1(\cdots) + s^0(\cdots) \big].$$

By the definition of the principal symbol, we then have

$$\sigma(h \mapsto \mathcal{L}_{(\delta G(h))^\sharp} g)(x, \xi)h = -\xi \otimes h(\xi^\sharp, \cdot) - h(\xi^\sharp, \cdot) \otimes \xi + (\xi \otimes \xi)\mathrm{tr}h,$$

and

$$\sigma(L)(x, \xi)h = |\xi|^2 h - \xi \otimes h(\xi^\sharp, \cdot) - h(\xi^\sharp, \cdot) \otimes \xi + (\xi \otimes \xi)\mathrm{tr}h.$$

Having computed the principal symbol, we can now check whether $\frac{\partial h}{\partial t} = L(h)$ is parabolic. This would require $\langle \sigma(L)(x, \xi)h, h \rangle > 0$ for all $(x, \xi) \in T^*\mathcal{M}$ with $\xi \neq 0$ and all $h \in \mathrm{Sym}^2 T_x^*\mathcal{M}$. However, setting $h = \xi \otimes \xi$ we see that

$$\sigma(L)(x, \xi)h = |\xi|^2 \xi \otimes \xi(1 - 1 - 1 + 1) = 0,$$

so $\frac{\partial h}{\partial t} = L(h)$ is *not* parabolic.

Remark 5.1.1. More generally, for $(x, \xi) \in T^*\mathcal{M}, \xi \neq 0$ and $\omega \in T_x^*\mathcal{M}$, we have $\sigma(L)(x, \xi)(\xi \otimes \omega + \omega \otimes \xi) = 0$.

Remark 5.1.2. This problem with lack of parabolicity arises from the diffeomorphism invariance of the equation, as discussed in [2, §5.C]. We resolve this problem by use of the so-called "DeTurck trick" [9].

5.2 Short-time existence and uniqueness: The DeTurck trick

Let $T \in \Gamma(\mathrm{Sym}^2 T^*\mathcal{M})$ be fixed, smooth, and positive definite. We also denote by T the invertible map $\Gamma(T^*\mathcal{M}) \to \Gamma(T^*\mathcal{M})$ which T induces, given a metric g. Define

$$P(g) = -2\mathrm{Ric}(g) + \mathcal{L}_{(T^{-1}\delta G(T))^\sharp} g. \tag{5.2.1}$$

As before, we write $Q := -2\mathrm{Ric}$. There are two steps to the DeTurck trick [9]:

STEP 1. Demonstrate that $\frac{\partial g}{\partial t} = P(g)$ is parabolic (and so admits solutions over some short time interval, given an initial metric);

STEP 2. Modify that solution to get a Ricci flow.

Step 1:
First we need to linearise $g \mapsto \mathcal{L}_{(T^{-1}\delta G(T))^\sharp} g$ to get its principal symbol. From Proposition 2.3.11, we know that if $g(t)$ is a family of metrics with $\frac{\partial g}{\partial t} = h$, then

$$\left(\frac{\partial}{\partial t}\delta G(T)\right) Z = -T((\delta G(h))^\sharp, Z) + \left\langle h, \nabla T(\cdot, \cdot, Z) - \frac{1}{2}\nabla_Z T\right\rangle,$$

and hence

$$\frac{\partial}{\partial t}T^{-1}\delta G(T) = -\delta G(h) + (\cdots)$$

where $(\cdots)$ consists of terms not involving any derivatives of h, whilst $\delta G(h)$ contains one derivative of h. It follows that

$$\frac{\partial}{\partial t}\mathcal{L}_{(T^{-1}\delta G(T))^\sharp} g = -\mathcal{L}_{(\delta G(T))^\sharp} g + (\cdots)$$

where the first term on the right-hand side contains a second derivative of h, and $(\cdots)$ denotes terms involving h and its first derivative. Thus, since $P(g) = Q(g) + \mathcal{L}_{(T^{-1}\delta G(T))^\sharp} g$,

$$
\begin{aligned}
DP(g)h &= [DQ(g) + D\mathcal{L}_{(T^{-1}\delta G(T))^\sharp} g]h \\
&= (\Delta_L h + \mathcal{L}_{(\delta G(h))^\sharp} g) - \mathcal{L}_{(\delta G(h))^\sharp} g + \text{lower order terms} \\
&= \Delta h + \text{lower order terms.}
\end{aligned}
$$

Therefore $\sigma(DP(g))(x, \xi)h = |\xi|^2 h$, and we see that $DP(g)$ is parabolic even in the special sense discussed in Remark 4.3.2). Consequently, the equation $\frac{\partial g}{\partial t} = P(g)$ is parabolic at any metric g, and given a smooth initial metric g_0, there exist $\varepsilon > 0$ and a solution $g(t)$ of

$$
\begin{cases}
\frac{\partial g}{\partial t} = P(g) & t \in [0, \varepsilon] \\
g(0) = g_0.
\end{cases}
\tag{5.2.2}
$$

Step 2:
From this solution to (5.2.2), we will construct a solution to the Ricci flow using the framework we used for Ricci solitons in Section 1.2.2:

(i) At each t, define a vector field $X := -(T^{-1}\delta G(T))^\sharp$.
(ii) Use this to generate a family of diffeomorphisms $\psi_t : \mathcal{M} \to \mathcal{M}$ with $\psi_0 = \text{id}$.
(iii) Define a new family of metrics $\hat{g}(t) := \psi_t^*(g(t))$.

(iv) By Proposition 1.2.1,

$$\frac{\partial \hat{g}}{\partial t} = \psi_t^* \left(\frac{\partial g}{\partial t} + \mathcal{L}_X g \right)$$

$$= \psi_t^* \left(-2\mathrm{Ric}(g) + \mathcal{L}_{(T^{-1}\delta G(T))^\sharp} g + \mathcal{L}_{-(T^{-1}\delta G(T))^\sharp} g \right)$$

$$= -2\psi_t^* \left(\mathrm{Ric}(g) \right)$$

$$= -2\mathrm{Ric}(\hat{g}),$$

and $\hat{g}(0) = \psi_0^*(g(0)) = g_0$ so $\hat{g}$ is the desired Ricci flow.

These two steps lead us to the following conclusion, first proved by Hamilton [19].

Theorem 5.2.1 (Short time existence). *Given a smooth metric g_0 on a closed manifold $\mathcal{M}$, there exist $\varepsilon > 0$ and a smooth family of metrics $g(t)$ for $t \in [0, \varepsilon]$ such that*

$$\begin{cases} \frac{\partial g}{\partial t} = -2\,\mathrm{Ric}(g) & t \in [0, \varepsilon] \\ g(0) = g_0. \end{cases} \tag{5.2.3}$$

The vector field $-(T^{-1}\delta G(T))^\#$ which we integrate to give the diffeomorphisms ψ_t may also be written $\tau_{g(t),T}(id)$, the *tension* of the identity map $(\mathcal{M}, g(t))$ to $(\mathcal{M}, T)$. (See [10] for a definition of the tension field, and [21, §6] for a discussion in this context.) The diffeomorphisms ψ_t may then be seen to represent a solution of the harmonic map flow [11], albeit with a time-dependent domain metric. Taking this viewpoint, if we are given two Ricci flows $g_1(t)$ and $g_2(t)$ on a time interval $[0, \varepsilon)$ which agree at time zero, then we may fix a metric T, and solve the harmonic map flow

$$\frac{\partial \psi^i}{\partial t} = \tau_{g_i(t),T}(\psi^i)$$

for the (unique) maps $\psi^i : \mathcal{M} \times [0, \epsilon) \to \mathcal{M}$ with $\psi^i(0) = id$, for some possibly smaller $\varepsilon > 0$ ($i = 1, 2$). By reducing $\varepsilon > 0$ if necessary, we may assume that each ψ^i remains a diffeomorphism on each time slice. But now, the pushforwards $(\psi^i)_*(g_i(t))$ again obey the parabolic equation (5.2.2) with the same initial metric, and must therefore agree over the whole time interval $[0, \varepsilon)$. We then see that the maps ψ^i are identical, and hence also the original Ricci flows $g_i(t)$. By arguing along these lines, one can even prove the following uniqueness:

Theorem 5.2.2 (Uniqueness of solutions (forwards and backwards)). *Suppose $g_1(t)$ and $g_2(t)$ are two Ricci flows on a closed manifold $\mathcal{M}$, for $t \in [0, \varepsilon], \varepsilon > 0$. If $g_1(s) = g_2(s)$ for some $s \in [0, \varepsilon]$, then $g_1(t) = g_2(t)$ for all $t \in [0, \varepsilon]$.*

Combining these two theorems, we can talk about *the* Ricci flow with initial metric g_0, on a *maximal* time interval $[0, T)$. In this situation, "maximal" means that either $T = \infty$, or that $T < \infty$ but there do not exist $\varepsilon > 0$ and a smooth Ricci flow $\hat{g}(t)$ for $t \in [0, T + \varepsilon)$ such that $\hat{g}(t) = g(t)$ for $t \in [0, T)$.

Remark 5.2.3. For existence and uniqueness results without the assumption that $\mathcal{M}$ is closed, see [36], [5] and [40].

5.3 Curvature blow-up at finite-time singularities

We can now also combine this theory with our global derivative estimates to prove:

Theorem 5.3.1 (Curvature blows up at a singularity). *If $\mathcal{M}$ is closed and $g(t)$ is a Ricci flow on a maximal time interval $[0, T)$ and $T < \infty$, then*

$$\sup_{\mathcal{M}} |\mathrm{Rm}|(\cdot, t) \to \infty \qquad (5.3.1)$$

as $t \uparrow T$.

The key ingredient is the following useful, elementary result, which prevents the metric from degenerating unless the Ricci curvature blows up.

Lemma 5.3.2 (Metric equivalence). *If $g(t)$ is a Ricci flow for $t \in [0, s]$ and $|\mathrm{Ric}| \le M$ on $\mathcal{M} \times [0, s]$, then*

$$e^{-2Mt} g(0) \le g(t) \le e^{2Mt} g(0) \qquad (5.3.2)$$

for all $t \in [0, s]$.

Remark 5.3.3. Of course, given two symmetric $(0, 2)$-tensors g_1 and g_2, we write $g_1 \ge g_2$ to mean that $g_1 - g_2$ is weakly positive definite.

Proof. Since $\frac{\partial}{\partial t} g(X, X) = -2 \mathrm{Ric}(X, X)$, we have

$$\left| \frac{\partial}{\partial t} g(X, X) \right| \le 2|\mathrm{Ric}| g(X, X),$$

and hence

$$\left| \frac{\partial}{\partial t} \ln g(X, X) \right| \le 2M,$$

for any nonzero tangent vector X. Therefore

$$\left| \ln \frac{g(t)(X, X)}{g(0)(X, X)} \right| \le 2Mt.$$

$\square$

Proof. (Theorem 5.3.1.)   We will deal with the contrapositive of this theorem, assuming instead that sup $|\text{Rm}|(\cdot, t) \not\to \infty$ as $t \uparrow T$, and proving that the Ricci flow can be extended to a larger time interval $[0, T + \varepsilon)$.

Note first that by Theorem 3.2.11, if sup $|\text{Rm}|(\cdot, t) \not\to \infty$, then there exists $M > 0$ such that $|\text{Rm}| \leq M$ for all $t \in [0, T)$. Indeed, this theorem tells us that if $|\text{Rm}| \leq M$ at time $t = T - \varepsilon$, then

$$|\text{Rm}| \leq \frac{M}{1 - CM[t - (T - \varepsilon)]}$$

for $t \in [T - \varepsilon, T]$.

We claim that in this case, $g(t)$ may be extended from being a smooth solution on $[0, T)$ to a smooth solution on $[0, T]$. We may then take $g(T)$ to be an 'initial' metric in our short-time existence theorem (Theorem 5.2.1) in order to extend the flow to a Ricci flow for $t \in [0, T + \varepsilon)$, contradicting the assumption that $[0, T)$ is a maximal time interval. (One should note here that the extended flow is smooth at $t = T$ – that is, $\frac{\partial^k g}{\partial t^k}$ exists at $t = T$ for $k \in \mathbb{N}$ – which can be seen by differentiating the equation for Ricci flow with respect to t, away from $t = T$, in order to write $\frac{\partial^k g}{\partial t^k}$ in terms of the curvature and its *spacial* derivatives.)

In order to prove our claim that $g(t)$ may be extended to a Ricci flow on $[0, T]$, we first note that by Lemma 5.3.2 and the boundedness of the curvature, the tensor $g(t)$ may be extended *continuously* to the time interval $[0, T]$, and the $g(T)$ which has been added will be a metric. (In particular, it will be positive definite.)

All that we need to show now, is that this extension is *smooth*. We will have achieved this if we can show that within a local coordinate chart, all space-time derivatives of g_{ij} are bounded on $[\frac{1}{2}T, T)$, for all indices i and j. By our global derivative estimates from Corollary 3.3.2, we have that

$$\left| \frac{\partial^l}{\partial t^l} \nabla^k \text{Rm} \right| \leq C = C(l, k, T, M, n)$$

for $t \in [\frac{1}{2}T, T)$. Working directly with the definition of Ric, and the Ricci flow equation (1.1.1), this may be turned into

$$\left| \frac{\partial^l}{\partial t^l} \nabla^k \text{Ric} \right| \leq C(l, k, T, M, n), \tag{5.3.3}$$

over the same time interval. Meanwhile, by using the consequence of (2.3.3) that

$$\left(\frac{\partial}{\partial t} \nabla - \nabla \frac{\partial}{\partial t} \right) A = A * \nabla \text{Rm}$$

for any t-dependent tensor field A, we can check that with respect to local coordinates $\{x^i\}$,

$$\left| \frac{\partial^l}{\partial t^l} \nabla^k \partial_i \right| \le C,$$

for each index i, where $\partial_i := \frac{\partial}{\partial x^i}$, and C is independent of $t \in [\frac{1}{2}T, T)$. By combining this with (5.3.3) we then find that for any indices a and b,

$$\left| \frac{\partial^l}{\partial t^l} D^\alpha R_{ab} \right| \le C \qquad (5.3.4)$$

for $t \in [\frac{1}{2}T, T)$, where $\alpha = (\alpha_1, \ldots, \alpha_p) \in (\{0\} \cup \mathbb{N})^p$ is any multi-index (that is, D^α is a combination of spatial partial derivatives, $D^\alpha = \frac{\partial^{|\alpha|}}{\partial x^{\alpha_1} \ldots \partial x^{\alpha_p}}$, with $|\alpha| := \alpha_1 + \ldots \alpha_p$).

Now by plugging the equation $\frac{\partial g_{ab}}{\partial t} = -2 R_{ab}$ into the inequality (5.3.4), we obtain the boundedness of

$$\left| \frac{\partial^l}{\partial t^l} D^\alpha g_{ab} \right|$$

for $l \ge 1$, any multi-index α, and any indices a and b, and by integrating with respect to time we find the same is true also for $l = 0$. $\square$

6

Ricci flow as a gradient flow

6.1 Gradient of total scalar curvature
and related functionals

Given a compact manifold with a Riemannian metric g, we define the total
scalar curvature by

$$E(g) = \int_{\mathcal{M}} R \, dV.$$

Let us consider the first variation of E under an arbitrary change of metric. As
before, we write $h = \frac{\partial g}{\partial t}$. Using Propositions 2.3.9 and 2.3.12, and keeping in
mind that the integral of the divergence of a 1-form must be zero, we compute

$$\frac{d}{dt} \int R \, dV = \int \frac{\partial R}{\partial t} dV + \int R \frac{1}{2} (\mathrm{tr} h) dV$$

$$= \int (-\langle \mathrm{Ric}, h \rangle + \delta^2 h - \Delta(\mathrm{tr} h)) dV + \int \frac{R}{2} \langle g, h \rangle dV$$

$$= \int \left\langle \frac{R}{2} g - \mathrm{Ric}, h \right\rangle dV.$$

By definition, the gradient of E is then given by

$$\nabla E(g) = \frac{R}{2} g - \mathrm{Ric}.$$

One might first consider the flow of metrics up the gradient of E – that is,
the flow $\frac{\partial g}{\partial t} = \frac{R}{2} g - \mathrm{Ric}$. Unfortunately, this flow cannot be expected to have
any solutions starting at an arbitrary metric g_0, in dimensions $n \geq 3$. This is
indicated by looking at how the scalar curvature would evolve under the flow;
by Proposition 2.3.9 and the identity $\delta \mathrm{Ric} + \frac{1}{2} dR = 0$ we would have

$$\frac{\partial R}{\partial t} = -\left(\frac{n}{2} - 1\right) \Delta R + |\mathrm{Ric}|^2 - \frac{1}{2} R^2, \qquad (6.1.1)$$

which is a *backwards* heat equation. Such equations do not typically admit solutions, although this issue is complicated in this situation by the fact that the Laplacian depends on the metric, and hence on the function R for which this is a PDE.

We should add that if only *conformal* deformations of metric are allowed, then the flow down the gradient is the so-called 'Yamabe flow,' also introduced by Hamilton. (See for example [43].)

6.2 The $\mathcal{F}$-functional

One of the early joys of the recent work of Perelman [31] is that Ricci flow *can* in fact be formulated as a gradient flow. To see one such formulation, we consider, for a fixed closed manifold $\mathcal{M}$, the following 'Fisher information' functional $\mathcal{F}$ on pairs (g, f) where g is a Riemannian metric, and f is a function $f : \mathcal{M} \to \mathbb{R}$:

$$\mathcal{F}(g, f) := \int (R + |\nabla f|^2) e^{-f} dV.$$

We need first to calculate the gradient of $\mathcal{F}$ under a smooth variation of g and f with $\frac{\partial g}{\partial t} = h$ (as usual) and $\frac{\partial f}{\partial t} = k$, for some function $k : \mathcal{M} \to \mathbb{R}$.

Proposition 6.2.1.

$$\frac{d}{dt}\mathcal{F}(g, f) = \int \langle -\mathrm{Ric} - \mathrm{Hess}(f), h \rangle e^{-f} dV$$
$$+ \int (2\Delta f - |\nabla f|^2 + R)\left(\frac{1}{2}\mathrm{tr}h - k\right) e^{-f} dV.$$

Proof. Keeping in mind that $\frac{\partial}{\partial t} g^{ij} = -h^{ij}$, we may calculate

$$\frac{\partial}{\partial t}|\nabla f|^2 = -h(\nabla f, \nabla f) + 2\langle \nabla k, \nabla f \rangle.$$

Therefore, using Propositions 2.3.9 and 2.3.12, we have

$$\frac{d}{dt}\mathcal{F}(g, f) = \int [-h(\nabla f, \nabla f) + 2\langle \nabla k, \nabla f \rangle - \langle \mathrm{Ric}, h \rangle + \delta^2 h - \Delta(\mathrm{tr}h)] e^{-f} dV$$
$$+ \int (R + |\nabla f|^2)\left[-k + \frac{1}{2}\mathrm{tr}h\right] e^{-f} dV.$$

Three of these terms may be usefully addressed by integrating by parts. First,

$$\int 2\langle \nabla k, \nabla f \rangle e^{-f} dV = \int -2k(\Delta f - |\nabla f|^2) e^{-f} dV.$$

Second, by Remark 2.1.1,

$$
\begin{aligned}
\int (\delta^2 h) e^{-f} dV &= \int \langle \delta h, d(e^{-f}) \rangle dV \\
&= \int \langle h, \nabla d(e^{-f}) \rangle dV \\
&= \int (h(\nabla f, \nabla f) - \langle \operatorname{Hess}(f), h \rangle) e^{-f} dV.
\end{aligned}
$$

Third,

$$
\int -\Delta(\operatorname{tr} h) e^{-f} dV = \int -(\operatorname{tr} h)\Delta(e^{-f}) dV = \int (\Delta f - |\nabla f|^2)(\operatorname{tr} h) e^{-f} dV.
$$

Combining these calculations, we find that

$$
\begin{aligned}
\frac{d}{dt}\mathcal{F}(g, f) &= \int \Bigg[-\langle \operatorname{Ric}, h \rangle - \langle \operatorname{Hess}(f), h \rangle + (\Delta f - |\nabla f|^2)(\operatorname{tr} h - 2k) \\
&\qquad + (R + |\nabla f|^2)\left(\frac{1}{2}\operatorname{tr} h - k\right) \Bigg] e^{-f} dV \\
&= \int \Bigg[\langle -\operatorname{Ric} - \operatorname{Hess}(f), h \rangle \\
&\qquad + (2\Delta f - |\nabla f|^2 + R)\left(\frac{1}{2}\operatorname{tr} h - k\right) \Bigg] e^{-f} dV
\end{aligned}
$$

(6.2.1)

□

6.3 The heat operator and its conjugate

We pause to define some notation. We denote the heat operator, acting on functions $f : \mathcal{M} \times [0, T] \to \mathbb{R}$ by

$$
\Box := \frac{\partial}{\partial t} - \Delta.
$$

We also need to define

$$
\Box^* := -\frac{\partial}{\partial t} - \Delta + R,
$$

which is conjugate to $\Box$ in the following sense.

Proposition 6.3.1. *If $g(t)$ is a Ricci flow for $t \in [0, T]$, and $v, w : \mathcal{M} \times [0, T] \to \mathbb{R}$ are smooth, then*

$$
\int_0^T \left(\int_{\mathcal{M}} (\Box v) w \, dV \right) dt = \left[\int_{\mathcal{M}} vw \, dV \right]_0^T + \int_0^T \left(\int_{\mathcal{M}} v(\Box^* w) dV \right) dt.
$$

(6.3.1)

Proof. Keeping in mind (2.5.7) we may calculate

$$\frac{d}{dt}\int vw\, dV = \int \left(\frac{\partial v}{\partial t}w + v\frac{\partial w}{\partial t} - Rvw\right) dV$$

$$= \int \left(((\Box v)w + (\Delta v)w) + (-v\Box^* w - v\Delta w + vwR) - Rvw\right) dV$$

$$= \int \left((\Box v)w - v(\Box^* w)\right) dV$$

$\square$

Remark 6.3.2. Several times we will need the special case of this calculation in which $v \equiv 1$, that is

$$\frac{d}{dt}\int w\, dV = -\int \Box^* w\, dV. \tag{6.3.2}$$

6.4 A gradient flow formulation

Let us now return to the discussion of Section 6.2, specialising to variations of g and f which preserve the distorted volume form $e^{-f}dV$. From Proposition 2.3.12, we see that

$$0 = \frac{\partial}{\partial t}(e^{-f}dV) = \left(\frac{1}{2}\text{tr}\frac{\partial g}{\partial t} - \frac{\partial f}{\partial t}\right)e^{-f}dV,$$

and so the evolution of f is determined by

$$\frac{\partial f}{\partial t} = \frac{1}{2}\text{tr}\frac{\partial g}{\partial t}$$

and the evolution of $\mathcal{F}$ proceeds according to

$$\frac{d}{dt}\mathcal{F}(g, f) = \int \left\langle -\text{Ric} - \text{Hess}(f), \frac{\partial g}{\partial t}\right\rangle e^{-f}dV.$$

From here, we can see that if we had a solution to the flow

$$\frac{\partial g}{\partial t} = -2(\text{Ric} + \text{Hess}(f)), \tag{6.4.1}$$

and the corresponding

$$\frac{\partial f}{\partial t} = -R - \Delta f, \tag{6.4.2}$$

then $\mathcal{F}$ would evolve under

$$\frac{d}{dt}\mathcal{F}(g, f) = 2\int |\text{Ric} + \text{Hess}(f)|^2 e^{-f}dV \geq 0. \tag{6.4.3}$$

In this case, defining

$$\omega := e^{-f} dV, \tag{6.4.4}$$

which would now be constant in time, we could view g as a gradient flow for the functional $g \mapsto \mathcal{F}(g, f)$, where f is determined from ω and g by (6.4.4).

The coupled system (6.4.1) and (6.4.2) looks at first glance to be somewhat alarming from a PDE point of view. Whilst (6.4.1) resembles the Ricci flow, which can be solved forwards in time for given initial data, (6.4.2) is a *backwards* heat equation and would not admit a solution for most prescribed initial data.

The trick to obtaining a solution of this system is to show that it is somehow equivalent to the decoupled system of equations

$$\frac{\partial g}{\partial t} = -2\text{Ric}, \tag{6.4.5}$$

$$\frac{\partial f}{\partial t} = -\Delta f + |\nabla f|^2 - R, \tag{6.4.6}$$

By the short-time existence of Theorem 5.2.1, (6.4.5) admits a smooth solution, for given initial data g_0, on some time interval $[0, T]$. As before, (6.4.6) is a backwards heat equation, and we cannot expect there to exist a solution achieving an arbitrary initial $f(0)$. Instead, for the given smooth $g(t)$, we prescribe *final* data $f(T)$ and solve backwards in time. Such a solution exists all the way down to time $t = 0$, for each smooth $f(T)$, as we can see by changing variables to

$$u(t) := e^{-f(t)}, \tag{6.4.7}$$

which satisfies the simple linear equation

$$\square^* u = 0, \tag{6.4.8}$$

which admits a positive solution backwards to time $t = 0$.

Given these $g(t)$ and $f(t)$, we set $X(t) := -\nabla f$ (which generates a unique family of diffeomorphisms ψ_t with $\psi_0 = identity$, say) so that $\mathcal{L}_X g = -2\text{Hess}(f)$ by (2.3.9). We fix $\sigma(t) \equiv 1$. Now when we define $\hat{g}(t)$ by (1.2.2), it will evolve, by (1.2.3), according to

$$\frac{\partial \hat{g}}{\partial t} = \psi_t^*[-2\text{Ric}(g) - 2\text{Hess}_g(f)],$$

where $\text{Hess}_g(f)$ is the Hessian of f with respect to the metric g. Keeping in mind that $\psi_t : (\mathcal{M}, \hat{g}) \to (\mathcal{M}, g)$ is an isometry, we may then write $\hat{f} := f \circ \psi_t$ to give

$$\frac{\partial \hat{g}}{\partial t} = -2 \left(\text{Ric}(\hat{g}) + \text{Hess}_{\hat{g}}(\hat{f}) \right). \tag{6.4.9}$$

The evolution of $\hat{f}$ is then found by the chain rule:

$$\frac{\partial \hat{f}}{\partial t}(x, t) = \frac{\partial f}{\partial t}(\psi_t(x), t) + X(f)(\psi_t(x), t)$$
$$= \left[(-\Delta_g f + |\nabla f|^2 - R_g) - |\nabla f|^2\right](\psi_t(x), t)$$
$$= \left[-\Delta_{\hat{g}} \hat{f} - R_{\hat{g}}\right](x, t),$$

where it has been necessary again to use subscripts to denote the metric being used. Thus we have a solution to

$$\frac{\partial \hat{f}}{\partial t} = -\Delta_{\hat{g}} \hat{f} - R_{\hat{g}} \tag{6.4.10}$$

coupled with (6.4.9), as desired. We have shown that solutions of (6.4.1) and (6.4.2) may be generated by pulling back solutions of (6.4.5) and (6.4.6) by an appropriate time-dependent diffeomorphism. One may also work the other way, finding solutions of (6.4.5) and (6.4.6) given solutions of (6.4.1) and (6.4.2).

Note that pulling back g and f by the same diffeomorphism leaves $\mathcal{F}$ invariant. In particular, $\mathcal{F}$ is monotonically increasing as g and f evolve under (6.4.5) and (6.4.6).

Let us take stock of what we have shown, in the following theorem.

Theorem 6.4.1. *For an arbitrary g_0, on a closed manifold $\mathcal{M}$, there exist $T > 0$, and an n-form $\omega > 0$ which is neither arbitrary nor unique, such that there exists a solution on the time interval $[0, T]$ to the (upwards) gradient flow for the functional*

$$\mathcal{F}^\omega(g) = \int (R + |\nabla f|^2)\omega,$$

where f is determined by the constraint $\omega = e^{-f}dV$, and we compute the gradient with respect to the inner product

$$\langle g, h \rangle_\omega = \frac{1}{2} \int \langle g, h \rangle \omega.$$

We are free to take T as large as we like, provided that the Ricci flow starting at g_0 exists for $t \in [0, T]$, although ω will depend on the T chosen.

This gradient flow is just the Ricci flow, modified by a time-dependent diffeomorphism. That is, there exists a smooth family of diffeomorphisms $\phi_t : \mathcal{M} \to \mathcal{M}$ such that $\phi_t^*(g(t))$ is the Ricci flow starting at g_0.

Remark 6.4.2. In terms of the construction above, we have

$$\omega = e^{-\hat{f}(0)}dV_{\hat{g}(0)} = e^{-f(0)}dV_{g_0},$$

where $f(0)$ is the result of the *smoothing* backwards flow (6.4.6) from the arbitrary $f(T)$, or alternatively,

$$\omega = e^{-\hat{f}(T)}dV_{\hat{g}(T)} = \psi_T^*(e^{-f(T)}dV_{g(T)}),$$

where $f(T)$ is arbitrary, but $g(T)$ is the result of the Ricci flow over the time interval $[0, T]$, and is thus necessarily very smooth.

Remark 6.4.3. By inspection of (6.4.3), we find that when we see $\mathcal{F}$ as a function of t, by defining g and f by (6.4.1) and (6.4.2), or equivalently by (6.4.5) and (6.4.5), if ever we had $\frac{d\mathcal{F}}{dt} = 0$ then we could deduce that Ric $+$ Hess(f) $= 0$, and by the discussion in Section 1.2.2, our Ricci flow would necessarily be a steady gradient soliton. In the present chapter, we are assuming that $\mathcal{M}$ is compact for simplicity, and that would force the flow to be Ricci flat, as one can see by analysing the scalar curvature near one of its minima. (See, for example, [7, Proposition 5.20].)

6.5 The classical entropy

So far in Chapter 6, our exposition has been an expansion of the original comments of Perelman [31]. It is enlightening also to emphasise the function u from (6.4.7) rather than f. If we specify $u(T)$ as a probability density at time T, and let it evolve backwards in time under (6.4.8), then $u(t)$ represents the probability density of a particle evolving under Brownian motion, backwards in time, on the evolving manifold. A basic building block for constructing interesting functionals for Ricci flow is then the classical, or 'Boltzmann-Shannon' entropy

$$N = - \int_{\mathcal{M}} u \ln u \, dV,$$

as considered in thermodynamics, information theory (Boltzmann, Shannon, Turing etc.) and elsewhere, which gives a measure of the concentration of the probability density. (Note that Brownian diffusion on a *fixed* manifold represents the gradient flow for this entropy on the space of probability densities endowed with the Wasserstein metric – see the discussion and references in [42].) In this framework, the information functional $\mathcal{F}$ coincides with the time derivative of N, up to a sign (see also [30]). By (2.5.7),

$$
\begin{aligned}
-\frac{dN}{dt} &= \int \left[(1 + \ln u) \frac{\partial u}{\partial t} - Ru \ln u \right] dV \\
&= -\int \left[(1 + \ln u)(\Delta u - Ru) + Ru \ln u \right] dV \\
&= \int \left(\frac{|\nabla u|^2}{u} + Ru \right) dV \\
&= \mathcal{F}.
\end{aligned}
\tag{6.5.1}
$$

It is sometimes convenient to take a renormalised version of the classical entropy

$$\tilde{N} = N - \frac{n}{2}(1 + \ln 4\pi (T - t)). \qquad (6.5.2)$$

With this adjustment, one can check that $\tilde{N} \to 0$ as $t \uparrow T$ when one takes u to be a solution of (6.4.8) which is a delta function at $t = T$. (In other words, u is a heat kernel for the conjugate heat equation – the probability density of the backwards Brownian path of a particle whose position is known at time $t = T$.) Indeed, this functional applied to the classical fundamental solution of the heat equation on Euclidean space would be zero. One could check that by virtue of the Harnack inequality of Li and Yau [28], the functional $\tilde{N}$ applied to a (backwards) heat kernel on a *fixed* manifold of *positive Ricci curvature* must be decreasing in $T - t$. For Ricci flow, that is true without any curvature hypothesis:

Lemma 6.5.1. *If $g(t)$ is a Ricci flow on a closed manifold $\mathcal{M}$, for $t \in [0, T]$, and $u : \mathcal{M} \times [0, T] \to (0, \infty)$ is a solution to $\Box^* u = 0$, then the functional $\tilde{N}$ defined by (6.5.2) is monotonically increasing in t.*

Proof. As observed above, we have

$$-\frac{d\tilde{N}}{dt} = \mathcal{F} - \frac{n}{2(T - t)},$$

so we are reduced to proving that

$$\mathcal{F} \le \frac{n}{2(T - t)}. \qquad (6.5.3)$$

The functional $\mathcal{F}$ evaluated as here on the Ricci flow $g(t)$ and the density u (or its corresponding $f := -\ln u$) is equal to $\mathcal{F}$ evaluated on the $\hat{g}$ and $\hat{f}$ of the previous section. We may then sharpen the monotonicity of $\mathcal{F}$, as expressed in (6.4.3) to

$$\frac{d\mathcal{F}}{dt} = 2 \int |\text{Ric} + \text{Hess}(f)|^2 e^{-f} dV \ge \frac{2}{n} \int |R + \Delta f|^2 e^{-f} dV$$

$$\ge \frac{2}{n} \left(\int (R + \Delta f) e^{-f} dV \right)^2 = \frac{2}{n} \mathcal{F}^2, \qquad (6.5.4)$$

where we have estimated the norm of the symmetric tensor $\text{Ric} + \text{Hess}(f)$ in terms of its trace, appealed to Cauchy-Schwarz, and integrated by parts, as in [31]. If ever (6.5.3) were violated, then the differential inequality (6.5.4) would force $\mathcal{F}(t)$ to blow up prior to the time T until which we are assuming its existence. $\square$

Essentially, this lemma says that the evolution of a manifold under Ricci flow will always balance with backwards Brownian motion to prevent diffusion faster than that on Euclidean space.

Currently the most useful entropy functional for Ricci flow is Perelman's $\mathcal{W}$-entropy which we shall meet in Chapter 8, and this too will arise from considering the entropy N.

6.6 The zeroth eigenvalue of $-4\Delta + R$

It is also useful to rewrite the functional $\mathcal{F}$ to act on pairs (g, ϕ) where $\phi := e^{-\frac{f}{2}}$. In this case, one can easily check that

$$\mathcal{F} = \int (4|\nabla\phi|^2 + R\phi^2)dV,$$

and hence that the zeroth eigenvalue of the operator $-4\Delta + R$ is given by

$$\lambda = \lambda(g) = \inf \left\{ \mathcal{F}(g, \phi) \mid \int \phi^2 dV = 1 \right\},$$

where we are using the same notation $\mathcal{F}$ even after the change of variables. By applying the direct method and elliptic regularity theory (see [15, §8.12]) we know that this infimum is attained by some smooth ϕ_{min} (for a given smooth g on a closed manifold) and also that after negating if necessary, we have $\phi_{min} > 0$. (Note that $|\phi_{min}|$ will also be a minimiser, and that by the Harnack inequality – see for example [15, Theorem 8.20] – we must then have $|\phi_{min}| > 0$.)

One consequence of this positivity is that the minimiser ϕ_{min} must be unique (up to a sign) or we would be able to take an appropriate linear combination of two distinct minimisers to give a new minimiser which would no longer be nonzero.

Another consequence is that we can write $\phi_{min} = e^{-\frac{f_{min}}{2}}$ for some function $f_{min} : \mathcal{M} \to \mathbb{R}$, and hence we see that

$$\lambda = \min \left\{ \mathcal{F}(g, f) \mid \int e^{-f} dV = 1 \right\}. \tag{6.6.1}$$

Theorem 6.6.1. *If $g(t)$ is a Ricci flow on a closed manifold $\mathcal{M}$, for $t \in [0, T]$, then the eigenvalue $\lambda(g(t))$ is a weakly increasing function on $[0, T]$.*

Proof. Pick arbitrary times a and b with $0 \leq a < b \leq T$, and define a function f_b to be the minimiser in (6.6.1) corresponding to the metric $g(b)$. Now if we define a time-dependent function $f(t)$ for $t \in [a, b]$ to solve (6.4.6) with final condition $f(b) = f_b$ (there is a unique smooth solution as in Section 6.4) then by the monotonicity of Section 6.4, we see that

$$\lambda(g(a)) \leq \mathcal{F}(g(a),\, f(a)) \leq \mathcal{F}(g(b),\, f(b)) = \lambda(g(b)).$$

$\square$

The monotonicity of such an eigenvalue may be considered – when $\lambda > 0$ – to be equivalent to the improvement of the best constant in an appropriate Poincaré inequality. In Chapter 8, we will see that the best constant in a certain log-Sobolev inequality is also improving.

7

Compactness of Riemannian manifolds and flows

Having shown in Section 5.3 that the curvature of a Ricci flow must blow up in magnitude at a singularity, we will now work towards a theory of 'blowing-up' whereby we rescale a flow more and more as we get closer and closer to a singularity, and hope that if we rescale by enough to keep the curvature under control, then we can pass to a limit of flows to give a Ricci flow which captures some of what is happening at the singularity. Refer back to Section 1.2.3 for a discussion of rescaling.

The first step in this direction is to pin down what it means for a sequence of flows, or indeed of manifolds, to converge. We then need some sort of compactness theorem to allow us to pass to a limit.

We will not give the rather long proofs of these compactness results, but refer the reader to [22].

7.1 Convergence and compactness of manifolds

It is reasonable to suggest that a sequence $\{g_i\}$ of Riemannian metrics on a manifold $\mathcal{M}$ should converge to a metric g when $g_i \to g$ as tensors. However, we would like a notion of convergence of Riemannian manifolds which is diffeomorphism invariant: it should not be affected if we modify each metric g_i by an i-dependent diffeomorphism. Once we have asked for such invariance, it is necessary to discuss convergence with respect to a point of reference on each manifold.

Definition 7.1.1 (Smooth, pointed "Cheeger-Gromov" convergence of manifolds). A sequence $(\mathcal{M}_i, g_i, p_i)$ of smooth, complete, *pointed* Riemannian manifolds (that is, Riemannian manifolds $(\mathcal{M}_i, g_i)$ and points $p_i \in \mathcal{M}_i$) is said to *converge* (smoothly) to the smooth, complete, pointed manifold $(\mathcal{M}, g, p)$ as $i \to \infty$ if there exist

(i) a sequence of compact sets $\Omega_i \subset \mathcal{M}$, exhausting $\mathcal{M}$ (that is, so that any compact set $K \subset \mathcal{M}$ satisfies $K \subset \Omega_i$ for sufficiently large i) with $p \in \text{int}(\Omega_i)$ for each i;
(ii) a sequence of smooth maps $\phi_i : \Omega_i \to \mathcal{M}_i$ which are diffeomorphic onto their image and satisfy $\phi_i(p) = p_i$ for all i;

such that,

$$\phi_i^* g_i \to g$$

smoothly as $i \to \infty$ in the sense that, for all compact sets $K \subset \mathcal{M}$, the tensor $\phi_i^* g_i - g$ and its covariant derivatives of all orders (with respect to any fixed background connection) each converge uniformly to zero on K.

To demonstrate why we need to include the points p_i in the above definition, consider the following example.

$(\mathcal{N}, h) =$

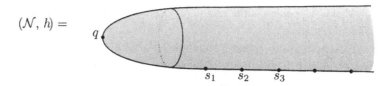

Suppose first that for every i, $(\mathcal{M}_i, g_i)$ is equal to the $(\mathcal{N}, h)$ as shown above. Then $(\mathcal{M}_i, g_i, q) \to (\mathcal{N}, h, q)$, but $(\mathcal{M}_i, g_i, s_i)$ converges to a cylinder.

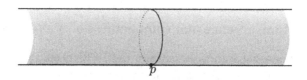

Remark 7.1.2. It is possible to have $\mathcal{M}_i$ compact for all i, but have the limit $\mathcal{M}$ non-compact. For example:

Two consequences of the convergence $(\mathcal{M}_i, g_i, p_i) \to (\mathcal{M}, g, p)$ are that

(i) for all $s > 0$ and $k \in \{0\} \cup \mathbb{N}$,

$$\sup_{i \in \mathbb{N}} \sup_{B_{g_i}(p_i, s)} \left| \nabla^k \mathrm{Rm}(g_i) \right| < \infty; \qquad (7.1.1)$$

(ii)

$$\inf_i \ \mathrm{inj}(\mathcal{M}_i, g_i, p_i) > 0, \qquad (7.1.2)$$

where $\mathrm{inj}(\mathcal{M}_i, g_i, p_i)$ denotes the injectivity radius of $(\mathcal{M}_i, g_i)$ at p_i.

In fact, conditions (i) and (ii) are *sufficient* for subconvergence. Various incarnations of the following result appear in, or can be derived from, papers of (for example) Greene and Wu [16], Fukaya [12] and Hamilton [22], all of which can be traced back to original ideas of Gromov [17] and Cheeger.

Theorem 7.1.3 (Compactness – manifolds). *Suppose that $(\mathcal{M}_i, g_i, p_i)$ is a sequence of complete, smooth, pointed Riemannian manifolds (all of dimension n) satisfying (7.1.1) and (7.1.2). Then there exists a complete, smooth, pointed Riemannian manifold $(\mathcal{M}, g, p)$ (of dimension n) such that after passing to some subsequence in i,*

$$(\mathcal{M}_i, g_i, p_i) \to (\mathcal{M}, g, p).$$

Remark 7.1.4. Most theorems of this type only require boundedness of $|\mathrm{Rm}|$ (or even $|\mathrm{Ric}|$) and then ask for weaker convergence. Our manifolds will always arise as time-slices of Ricci flows and so we may as well assume $|\nabla^k \mathrm{Rm}|$ bounds by our derivative estimates from Theorem 3.3.1.

We clearly need some curvature control, otherwise we may have:

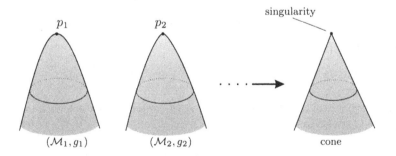

p_1 p_2 singularity

$(\mathcal{M}_1, g_1)$ $(\mathcal{M}_2, g_2)$ cone

There are other notions of convergence which can handle this type of limit. For example, Gromov-Hausdorff convergence allows us to take limits of metric spaces. (See, for example [3].)

Remark 7.1.5. The uniform positive lower bound on the injectivity radius is also necessary, since otherwise we could have, for example, degeneration of cylinders:

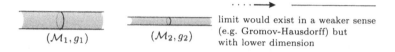

		limit would exist in a weaker sense
$(\mathcal{M}_1, g_1)$	$(\mathcal{M}_2, g_2)$	(e.g. Gromov-Hausdorff) but with lower dimension

Remark 7.1.6. Given curvature bounds as in (7.1.1), the injectivity radius lower bound at p_i implies a positive lower bound for the injectivity radius at other points $q \in \mathcal{M}_i$ in terms of the distance from p_i to q, and the curvature bounds, as discussed in [22], say.

Remark 7.1.7. In these notes, whenever we wish to apply Theorem 7.1.3, we will know not only (7.1.1) but also that for all $k \in \{0\} \cup \mathbb{N}$,

$$\sup_i \sup_{x \in \mathcal{M}} \left| \nabla^k \mathrm{Rm}(g_i) \right| < \infty.$$

7.2 Convergence and compactness of flows

One can derive, from the compactness theorem for manifolds (Theorem 7.1.3) a compactness theorem for Ricci flows.

Definition 7.2.1. Let $(\mathcal{M}_i, g_i(t))$ be a sequence of smooth families of complete Riemannian manifolds for $t \in (a, b)$ where $-\infty \leq a < 0 < b \leq \infty$. Let $p_i \in \mathcal{M}_i$ for each i. Let $(\mathcal{M}, g(t))$ be a smooth family of complete Riemannian manifolds for $t \in (a, b)$ and let $p \in \mathcal{M}$. We say that

$$(\mathcal{M}_i, g_i(t), p_i) \to (\mathcal{M}, g(t), p)$$

as $i \to \infty$ if there exist

(i) a sequence of compact $\Omega_i \subset \mathcal{M}$ exhausting $\mathcal{M}$ and satisfying $p \in \mathrm{int}(\Omega_i)$ for each i;

(ii) a sequence of smooth maps $\phi_i : \Omega_i \to \mathcal{M}_i$, diffeomorphic onto their image, and with $\phi_i(p) = p_i$;

such that

$$\phi_i^* g_i(t) \to g(t)$$

as $i \to \infty$ in the sense that $\phi_i^* g_i(t) - g(t)$ and its derivatives of every order (with respect to time as well as covariant space derivatives with respect to any

fixed background connection) converge uniformly to zero on every compact
subset of $\mathcal{M} \times (a, b)$.

Remark 7.2.2. It also makes sense to talk about convergence on (for example)
the time interval $(-\infty, 0)$, even when flows are defined only for $(-T_i, 0)$ with
$T_i \to \infty$.

Hamilton then proves the following result [22], based on the core Theorem
7.1.3.

Theorem 7.2.3 (Compactness of Ricci flows). *Let $\mathcal{M}_i$ be a sequence of
manifolds of dimension n, and let $p_i \in \mathcal{M}_i$ for each i. Suppose that $g_i(t)$ is
a sequence of complete Ricci flows on $\mathcal{M}_i$ for $t \in (a, b)$, where $-\infty \le a <
0 < b \le \infty$. Suppose that*

(*i*)
$$\sup_i \sup_{x \in \mathcal{M}_i, \, t \in (a,b)} \left| \mathrm{Rm}(g_i(t)) \right|(x) < \infty; \quad and$$

(*ii*)
$$\inf_i \mathrm{inj}(\mathcal{M}_i, g_i(0), p_i) > 0.$$

*Then there exist a manifold $\mathcal{M}$ of dimension n, a complete Ricci flow $g(t)$ on
$\mathcal{M}$ for $t \in (a, b)$, and a point $p \in \mathcal{M}$ such that, after passing to a subsequence
in i,*

$$(\mathcal{M}_i, g_i(t), p_i) \to (\mathcal{M}, g(t), p),$$

as $i \to \infty$.

7.3 Blowing up at singularities I

As we have mentioned before, one application of the compactness of Ricci
flows in Theorem 7.2.3 that we have in mind, is to analyse rescalings of Ricci
flows near their singularities. Let $(\mathcal{M}, g(t))$ be a Ricci flow with $\mathcal{M}$ closed, on
the maximal time interval $[0, T)$ – as defined at the end of Section 5.2 – with
$T < \infty$. By Theorem 5.3.1 we know that

$$\sup_{\mathcal{M}} |\mathrm{Rm}|(\cdot, t) \to \infty$$

as $t \uparrow T$. Let us choose points $p_i \in \mathcal{M}$ and times $t_i \uparrow T$ such that

$$|\mathrm{Rm}|(p_i, t_i) = \sup_{x \in \mathcal{M}, \, t \in [0, t_i]} |\mathrm{Rm}|(x, t),$$

by, for example, picking (p_i, t_i) to maximise $|\text{Rm}|$ over $\mathcal{M} \times [0, T - \frac{1}{i}]$. Notice in particular that $|\text{Rm}|(p_i, t_i) \to \infty$ as $i \to \infty$. Define rescaled (and translated) flows $g_i(t)$ by

$$g_i(t) = |\text{Rm}|(p_i, t_i)\, g\left(t_i + \frac{t}{|\text{Rm}|(p_i, t_i)}\right).$$

By the discussion in Section 1.2.3, $(\mathcal{M}, g_i(t))$ is a Ricci flow on the time interval $[-t_i|\text{Rm}|(p_i, t_i), (T - t_i)|\text{Rm}|(p_i, t_i))$.

Moreover, for each i, $|\text{Rm}(g_i(0))|(p_i) = 1$ and for $t \leq 0$,

$$\sup_{\mathcal{M}} |\text{Rm}(g_i(t))| \leq 1.$$

Furthermore, by Theorem 3.2.11, for $t > 0$,

$$\sup_{\mathcal{M}} |\text{Rm}(g_i(t))| \leq \frac{1}{1 - C(n)t}.$$

Therefore, for all $a < 0$ and some $b = b(n) > 0$, $g_i(t)$ is defined for $t \in (a, b)$ and

$$\sup_i \sup_{\mathcal{M} \times (a,b)} |\text{Rm}(g_i(t))| < \infty.$$

By Theorem 7.2.3, we can pass to a subsequence in i, and get convergence $(\mathcal{M}, g_i(t), p_i) \to (\mathcal{N}, \hat{g}(t), p_\infty)$ to a "singularity model" Ricci flow $(\mathcal{N}, \hat{g}(t))$, *provided* that we can establish the injectivity radius estimate

$$\inf_i \text{inj}(\mathcal{M}, g_i(0), p_i) > 0. \tag{7.3.1}$$

Example 7.3.1. As an example of what we have in mind, recall the possibility of a 'neck pinch' as discussed in Section 1.3.2. The limit of the rescaled flows should, in principle, be a shrinking cylinder.

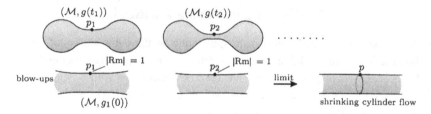

The missing step for us at this stage is to get the lower bound on the injectivity radius stated in (7.3.1). Historically, this has been a major difficulty, except in some special cases. However, as we will see in the next chapter, this issue has been resolved recently by Perelman.

8

Perelman's $\mathcal{W}$ entropy functional

In this chapter, we describe an entropy functional which is more useful than the $\mathcal{F}$ functional described in Section 6.2. Our main application of this will be to prove the missing injectivity radius estimate discussed in the previous chapter.

8.1 Definition, motivation and basic properties

Consider now the functional

$$\mathcal{W}(g, f, \tau) := \int \left[\tau(R + |\nabla f|^2) + f - n \right] u \, dV,$$

where g is a metric, $f : \mathcal{M} \to \mathbb{R}$ is smooth, $\tau > 0$ is a scale parameter, and u is defined by

$$u := (4\pi\tau)^{-\frac{n}{2}} e^{-f}. \tag{8.1.1}$$

Definition 8.1.1. The arguments g, f and τ are called *compatible* if

$$\int_{\mathcal{M}} u \, dV \equiv \int_{\mathcal{M}} \frac{e^{-f}}{(4\pi\tau)^{\frac{n}{2}}} \, dV = 1. \tag{8.1.2}$$

The scale factor τ is for convenience, in that it could be absorbed into the other two arguments of $\mathcal{W}$:

Proposition 8.1.2. *Under the transformation* $(g, f, \tau) \to (\tau^{-1}g, f, 1)$, *compatibility is preserved, as is the functional* $\mathcal{W}$:

$$\mathcal{W}(g, f, \tau) = \mathcal{W}(\tau^{-1}g, f, 1).$$

One motivation for the $\mathcal{W}$-entropy comes from the log-Sobolev inequality of L. Gross [18], as we now describe. We will later show how $\mathcal{W}$ arises from considering the classical entropy from Section 6.5.

71

On $\mathbb{R}^n$, the *Gaussian measure* $d\mu$ is defined in terms of the Lebesgue measure dx by

$$d\mu = (2\pi)^{-\frac{n}{2}} e^{-\frac{|x|^2}{2}} dx,$$

the normalisation being chosen so that

$$\int_{\mathbb{R}^n} d\mu = 1.$$

Theorem 8.1.3 (L.Gross). *If $v : \mathbb{R}^n \to \mathbb{R}$ is, say, smooth and satisfies v, $|\nabla v| \in L^2(d\mu)$ then*

$$\int v^2 \ln |v| \, d\mu \le \int |\nabla v|^2 \, d\mu + \left(\int v^2 \, d\mu \right) \ln \left(\int v^2 \, d\mu \right)^{\frac{1}{2}}. \qquad (8.1.3)$$

Remark 8.1.4.   We have equality in (8.1.3) if and only if v is constant.

Remark 8.1.5.   There is no constant in this inequality which depends on the dimension n.

Remark 8.1.6.   If we choose v so that $\int v^2 \, d\mu = 1$, then the inequality becomes

$$\int v^2 \ln |v| \, d\mu \le \int |\nabla v|^2 \, d\mu.$$

Let us consider the consequences of this inequality for the functional $\mathcal{W}$ in the case that $(\mathcal{M}, g)$ is Euclidean space.

Lemma 8.1.7.   *Let g denote the standard flat metric on $\mathcal{M} = \mathbb{R}^n$. If f and τ are compatible with g, then*

$$\mathcal{W}(g, f, \tau) \ge 0 \qquad (8.1.4)$$

with equality if and only if $f(x) \equiv \frac{|x|^2}{4\tau}$.

Proof.   For now, set $\tau = \frac{1}{2}$ and let $f : \mathbb{R}^n \to \mathbb{R}$ be compatible with g and τ, which in this situation means that

$$\int_{\mathbb{R}^n} \frac{e^{-f}}{(2\pi)^{\frac{n}{2}}} \, dx = 1.$$

If we define $v = e^{\frac{|x|^2}{4} - \frac{f}{2}}$, we have

$$v^2 \, d\mu = e^{\frac{|x|^2}{2} - f} (2\pi)^{-\frac{n}{2}} e^{-\frac{|x|^2}{2}} \, dx = (2\pi)^{-\frac{n}{2}} e^{-f} \, dx$$

and so $\int_{\mathbb{R}^n} v^2 \, d\mu = 1$. Therefore, by the log-Sobolev inequality,

$$\int v^2 \ln |v| \, d\mu \le \int |\nabla v|^2 \, d\mu.$$

We wish to convert from v to f in this inequality. For the left-hand side, we have

$$\int v^2 \ln |v| \, d\mu = \int e^{\frac{|x|^2}{2} - f} \left(\frac{|x|^2}{4} - \frac{f}{2} \right) (2\pi)^{-\frac{n}{2}} e^{-\frac{|x|^2}{2}} \, dx$$

$$= \int \left(\frac{|x|^2}{4} - \frac{f}{2} \right) \frac{e^{-f}}{(2\pi)^{\frac{n}{2}}} \, dx.$$

For the right-hand side, first note that

$$\nabla v = \left(\frac{x}{2} - \frac{\nabla f}{2} \right) e^{\frac{|x|^2}{4} - \frac{f}{2}}$$

which gives us

$$|\nabla v|^2 = \left(\frac{|x|^2}{4} - \frac{x.\nabla f}{2} + \frac{|\nabla f|^2}{4} \right) e^{\frac{|x|^2}{2} - f},$$

and thus

$$|\nabla v|^2 \, d\mu = \left(\frac{|x|^2}{4} - \frac{x.\nabla f}{2} + \frac{|\nabla f|^2}{4} \right) \frac{e^{-f}}{(2\pi)^{\frac{n}{2}}} \, dx.$$

Before integrating this, we use the fact that $\nabla.x = n$ to give the integration-by-parts formula

$$- \int \frac{x.\nabla f}{2} \frac{e^{-f}}{(2\pi)^{\frac{n}{2}}} \, dx = \frac{1}{2} \int x.\nabla(e^{-f}) \frac{dx}{(2\pi)^{\frac{n}{2}}} = -\frac{n}{2} \int \frac{e^{-f}}{(2\pi)^{\frac{n}{2}}} \, dx.$$

Therefore,

$$\int |\nabla v|^2 \, d\mu = \int \left(\frac{|x|^2}{4} - \frac{n}{2} + \frac{|\nabla f|^2}{4} \right) \frac{e^{-f}}{(2\pi)^{\frac{n}{2}}} \, dx,$$

and the log-Sobolev inequality (8.1.3) becomes

$$\int \left(\frac{|x|^2}{4} - \frac{f}{2} \right) \frac{e^{-f}}{(2\pi)^{\frac{n}{2}}} \, dx \leq \int \left(\frac{|x|^2}{4} - \frac{n}{2} + \frac{|\nabla f|^2}{4} \right) \frac{e^{-f}}{(2\pi)^{\frac{n}{2}}} \, dx$$

or equivalently

$$\mathcal{W} \left(g, f, \frac{1}{2} \right) := \int \left[\frac{1}{2} |\nabla f|^2 + f - n \right] \frac{e^{-f}}{(2\pi)^{\frac{n}{2}}} \, dx \geq 0,$$

for this special case where $(\mathcal{M}, g)$ is Euclidean n-space, with equality if and only if $f(x) \equiv \frac{|x|^2}{2}$ by Remark 8.1.4. By the scale invariance $\mathcal{W}(g, f, \tau) = \mathcal{W}(\frac{1}{2\tau} g, f, \frac{1}{2})$, the fact that compatibility is preserved under the transformation $(g, f, \tau) \mapsto (\frac{1}{2\tau} g, f, \frac{1}{2})$ and also the fact that $(\mathbb{R}^n, g)$ is preserved under homothetic scaling, we conclude that

$$\mathcal{W}(g, f, \tau) \geq 0, \tag{8.1.5}$$

for any f, τ compatible with g, with equality if and only if $f(x) \equiv \frac{|x|^2}{4\tau}$. $\quad\square$

In contrast to the situation for the $\mathcal{F}$-functional in Section 6.6, one must take a little care to show that the $\mathcal{W}$-functional is bounded below and that a minimising compatible f exists, given g and τ.

Lemma 8.1.8. *For any smooth Riemannian metric g on a closed manifold $\mathcal{M}$, and $\tau > 0$, the infimum of $\mathcal{W}(g, f, \tau)$ over all compatible f (as defined in Definition 8.1.1) is attained by a smooth compatible f.*

Defining μ as this infimum,

$$\mu(g, \tau) := \inf_f \mathcal{W}(g, f, \tau), \tag{8.1.6}$$

the function $\mu(g, \tau)$ is bounded below as τ varies within any finite interval $(0, \tau_0]$.

Given this lemma, we may define, for $\tau_0 > 0$,

$$\nu(g, \tau_0) := \inf_{\tau \in (0, \tau_0]} \mu(g, \tau) > -\infty. \tag{8.1.7}$$

The starting point for the proof of this lemma is the change of variables $\phi := e^{-\frac{f}{2}}$ that we used in Section 6.6. Abusing notation for $\mathcal{W}$, we then have

$$\mathcal{W}(g, \phi, \tau) = (4\pi\tau)^{-\frac{n}{2}} \int \left[\tau(4|\nabla\phi|^2 + R\phi^2) - 2\phi^2 \ln\phi - n\phi^2 \right] dV, \tag{8.1.8}$$

and the compatibility constraint (8.1.2) becomes

$$(4\pi\tau)^{-\frac{n}{2}} \int \phi^2 dV = 1. \tag{8.1.9}$$

These expressions have the benefit of making perfect sense when ϕ is merely weakly (rather than strictly) positive, and by approximation,

$$\inf_f \mathcal{W}(g, f, \tau) = \inf_\phi \mathcal{W}(g, \phi, \tau), \tag{8.1.10}$$

where the infima are taken over $f : \mathcal{M} \to \mathbb{R}$ and $\phi : \mathcal{M} \to [0, \infty)$ compatible with g and τ, and we are abusing notation for $\mathcal{W}$ as usual.

We will use this change of variables to prove the existence of a minimising compatible ϕ, given g and τ. One can argue that this minimiser will be strictly positive (c.f. Section 6.6, a proof being available in [34]) and hence one has the existence of a minimising f as asserted in Lemma 8.1.8.

We will not prove the asserted boundedness of μ, which is from [31], since it requires a little more analysis than we are assuming.

Proof. (Parts of Lemma 8.1.8.)   We limit ourselves to proving that the functional $\mathcal{W}$ is bounded below, for given g and τ, and that a minimising sequence of compatible positive functions ϕ must be bounded in $W^{1,2}$, and hence that the direct method applies to give a minimiser. (See [15, §8.12].)

By the invariance observed in Proposition 8.1.2, we may scale g and τ simultaneously and assume that $\tau = \frac{1}{4\pi}$, say, which slightly simplifies (8.1.8) and (8.1.9).

It suffices to prove that

$$
\begin{aligned}
\mathcal{W}(\phi) &:= \int \left[\frac{1}{4\pi}(4|\nabla\phi|^2 + R\phi^2) - 2\phi^2 \ln\phi - n\phi^2 \right] dV \\
&\geq \frac{1}{2\pi} \int |\nabla\phi|^2 dV - C(g, n),
\end{aligned}
\tag{8.1.11}
$$

for any positive ϕ satisfying

$$
\int \phi^2 dV = 1,
\tag{8.1.12}
$$

for some constant $C(g, n)$.

To achieve this, we split up the expression (8.1.11) for $\mathcal{W}(\phi)$. First, by (8.1.12) we clearly have

$$
\int \left[\frac{1}{4\pi}(R\phi^2) - n\phi^2 \right] dV \geq -C(g, n),
\tag{8.1.13}
$$

since our manifold is closed, so R is bounded.

Meanwhile, we may write

$$
\begin{aligned}
\int \phi^2 \ln\phi \, dV &= \frac{n-2}{4} \int \phi^2 \ln\phi^{\frac{4}{n-2}} dV \\
&= \frac{n-2}{4} \int G(\sigma) d\mu
\end{aligned}
\tag{8.1.14}
$$

where $\sigma : \mathcal{M} \to (0, \infty)$ is defined by $\sigma := \phi^{\frac{4}{n-2}}$, the function $G : (0, \infty) \to \mathbb{R}$ defined by $G(y) := \ln y$ is concave, and the measure $d\mu := \phi^2 dV$ has unit total mass. We may apply Jensen's inequality

$$
\int G(\sigma) d\mu \leq G\left(\int \sigma d\mu \right),
$$

(see for example [41, p.11]) to tell us that

$$
\int \phi^2 \ln\phi \, dV \leq \frac{n-2}{4} \ln\left[\int \phi^{2+\frac{4}{n-2}} dV \right] = \frac{n}{4} \ln\left[\int \phi^{\frac{2n}{n-2}} dV \right]^{\frac{n-2}{n}}.
$$

We continue by applying the Sobolev inequality (see [15])

$$
\left(\int \phi^{\frac{2n}{n-2}} dv \right)^{\frac{n-2}{2n}} \leq C(g, n) \left[\int \phi^2 dV + \int |\nabla\phi|^2 dV \right]^{\frac{1}{2}},
$$

to give

$$\int \phi^2 \ln \phi \, dV \leq C(g,n) + \frac{n}{4} \ln \left[1 + \int |\nabla \phi|^2 dV \right] \leq \frac{1}{2\pi} \int |\nabla \phi|^2 dV + C(g,n),$$

where the constant $C(g,n)$ is permitted to change at each step.

Combining with (8.1.13), we conclude that

$$\mathcal{W}(\phi) \geq \frac{1}{2\pi} \int |\nabla \phi|^2 dV - C(g,n),$$

as required in (8.1.11).   □

8.2 Monotonicity of $\mathcal{W}$

In this section, we would like to show that $\mathcal{W}$, like $\mathcal{F}$, is increasing under the Ricci flow when f and τ are made to evolve appropriately.

Proposition 8.2.1. *If M is closed, and g, f and τ evolve according to*

$$\begin{cases} \frac{\partial g}{\partial t} = -2\mathrm{Ric} \\ \frac{d\tau}{dt} = -1 \\ \frac{\partial f}{\partial t} = -\Delta f + |\nabla f|^2 - R + \frac{n}{2\tau} \end{cases} \tag{8.2.1}$$

then the functional $\mathcal{W}$ increases according to

$$\frac{d}{dt}\mathcal{W}(g,f,\tau) = 2\tau \int \left| \mathrm{Ric} + \mathrm{Hess}(f) - \frac{g}{2\tau} \right|^2 u \, dV \geq 0. \tag{8.2.2}$$

Remark 8.2.2. Under the evolution of the previous proposition, u satisfies

$$\square^* u = 0,$$

where $\square^*$ has been defined in Section 6.3. One consequence is that the compatibility constraint (8.1.2) is preserved under the evolution in (8.2.1), because by (6.3.2)

$$\frac{d}{dt}\int u \, dV = -\int \square^* u \, dV = 0.$$

Indeed, as we discussed in Section 6.5, u is again evolving as the probability density corresponding to Brownian motion backwards in time. More generally, by Proposition 6.3.1, if h is a solution of the heat equation $\square h = 0$, then

$$\int u h \, dV$$

is constant in time.

Remark 8.2.3. If the time derivative of $\mathcal{W}$ in (8.2.2) ever fails to be *strictly* positive, then we clearly have $\text{Ric} + \text{Hess}(f) - \frac{g}{2\tau} = 0$, and our flow is a shrinking gradient soliton, as discussed in Section 1.2.2.

Remark 8.2.4. The $\mathcal{W}$-functional applied to this backwards Brownian diffusion on a Ricci flow also arises via the (renormalised) classical entropy $\tilde{N}$ discussed in Section 6.5. Indeed, if we set $\tau = T - t$, one may calculate (compare with [31] and [30]) that

$$\mathcal{W}(t) = -\frac{d}{dt}(\tau \tilde{N}).$$

Remark 8.2.5. As for the existence of a solution to the system (8.2.1), the story is more or less the same as in Section 6.4; once we have a Ricci flow on a time interval $[0, T]$ (for example from the short time existence result) we may set $\tau = T + C - t$ for some $C > 0$, and solve for f for prescribed *final* data $f(T)$ by considering the linear equation satisfied by u in Remark 8.2.2.

The following result can be considered to be a *local* version of Proposition 8.2.1.

Proposition 8.2.6. *Suppose g, f and τ evolve as in Proposition 8.2.1, and u is defined as in (8.1.1). Then the function v defined by*

$$v := \left[\tau(2\Delta f - |\nabla f|^2 + R) + f - n\right]u \qquad (8.2.3)$$

satisfies

$$\square^* v = -2\tau|\text{Ric} + \text{Hess}(f) - \frac{g}{2\tau}|^2 u.$$

Remark 8.2.7. Proposition 8.2.1 follows immediately from Proposition 8.2.6 because by the integration by parts formula

$$\int \Delta u \, dV \equiv \int (\Delta f - |\nabla f|^2)u \, dV = 0,$$

we have

$$\mathcal{W} = \int_M v \, dV,$$

and so by (6.3.2)

$$\frac{d}{dt}\mathcal{W} = -\int_M \square^* v \, dV.$$

Proof. (Proposition 8.2.6.) Splitting v as a product, we find that

$$\square^* v = \square^*\left(\frac{v}{u}u\right) = \frac{v}{u}\square^* u - u\left(\frac{\partial}{\partial t} + \Delta\right)\left(\frac{v}{u}\right) - 2\left\langle\nabla\left(\frac{v}{u}\right), \nabla u\right\rangle,$$

and since $\Box^* u = 0$, we have

$$u^{-1}\Box^* v = -\left(\frac{\partial}{\partial t} + \Delta\right)\left(\frac{v}{u}\right) - 2u^{-1}\left\langle \nabla\left(\frac{v}{u}\right), \nabla u\right\rangle. \qquad (8.2.4)$$

For the first of the terms on the right-hand side, we compute

$$-\left(\frac{\partial}{\partial t} + \Delta\right)\left(\frac{v}{u}\right) = -\left(\frac{\partial}{\partial t} + \Delta\right)[\tau(2\Delta f - |\nabla f|^2 + R) + f - n]$$

$$= (2\Delta f - |\nabla f|^2 + R)$$

$$- \tau\left(\frac{\partial}{\partial t} + \Delta\right)(2\Delta f - |\nabla f|^2 + R) - \left(\frac{\partial}{\partial t} + \Delta\right)f,$$

and using the evolution equation for f in (8.2.1) on the final term, this reduces to

$$-\left(\frac{\partial}{\partial t} + \Delta\right)\left(\frac{v}{u}\right) = 2\Delta f - 2|\nabla f|^2 + 2R - \frac{n}{2\tau}$$

$$- \tau\left(\frac{\partial}{\partial t} + \Delta\right)(2\Delta f - |\nabla f|^2 + R). \qquad (8.2.5)$$

We deal with each part of the final term individually:

By Proposition 2.5.6 and the evolution equation for f,

$$\left(\frac{\partial}{\partial t} + \Delta\right)\Delta f = \Delta\frac{\partial f}{\partial t} + 2\langle \text{Ric}, \text{Hess}(f)\rangle + \Delta^2 f$$

$$= 2\langle \text{Ric}, \text{Hess}(f)\rangle + \Delta\left(|\nabla f|^2 - R + \frac{n}{2\tau}\right) \qquad (8.2.6)$$

Using also the evolution equation for g,

$$\left(\frac{\partial}{\partial t} + \Delta\right)|\nabla f|^2 = 2\text{Ric}(\nabla f, \nabla f) + 2\left\langle \nabla f, \nabla\frac{\partial f}{\partial t}\right\rangle + \Delta|\nabla f|^2$$

$$= 2\text{Ric}(\nabla f, \nabla f) + 2\langle \nabla f, \nabla(-\Delta f + |\nabla f|^2 - R)\rangle + \Delta|\nabla f|^2,$$

and by Proposition 2.5.4,

$$\left(\frac{\partial}{\partial t} + \Delta\right)R = 2\Delta R + 2|\text{Ric}|^2.$$

Combining these three expressions gives

$$\left(\frac{\partial}{\partial t} + \Delta\right)(2\Delta f - |\nabla f|^2 + R) = 4\langle \text{Ric}, \text{Hess}(f)\rangle + \Delta|\nabla f|^2 - 2\text{Ric}(\nabla f, \nabla f)$$

$$- 2\langle \nabla f, \nabla(-\Delta f + |\nabla f|^2 - R)\rangle + 2|\text{Ric}|^2. \qquad (8.2.7)$$

Also,

$$\left\langle \nabla \frac{v}{u}, \nabla u \right\rangle = \langle \tau \nabla (2\Delta f - |\nabla f|^2 + R) + \nabla f, -u\nabla f \rangle$$

so

$$-2u^{-1}\left\langle \nabla \frac{v}{u}, \nabla u \right\rangle = 2\tau \langle \nabla(2\Delta f - |\nabla f|^2 + R), \nabla f \rangle + 2|\nabla f|^2 \qquad (8.2.8)$$

Combining (8.2.4), (8.2.5), (8.2.7) and (8.2.8) we find that

$$u^{-1}\Box^* v = \frac{1}{\tau}\left(-\frac{n}{2}\right) + (2\Delta f + 2R)$$
$$+ \tau \Big(-4\langle \text{Ric}, \text{Hess}(f) \rangle - 2|\text{Ric}|^2$$
$$+ \big[-\Delta |\nabla f|^2 + 2\text{Ric}(\nabla f, \nabla f) + 2\langle \nabla f, \nabla \Delta f \rangle \big] \Big).$$

The three terms within square brackets simplify to $-2|\text{Hess}(f)|^2$ by the Bochner formula, so

$$u^{-1}\Box^* v = -\frac{n}{2\tau} + 2\langle \text{Ric} + \text{Hess}(f), g \rangle - 2\tau |\text{Ric} + \text{Hess}(f)|^2$$
$$= -2\tau \left| \text{Ric} + \text{Hess}(f) - \frac{g}{2\tau} \right|^2$$

$$\Box$$

8.3 No local volume collapse where curvature is controlled

Given a complete Riemannian manifold $(\mathcal{M}, g)$, a point $p \in \mathcal{M}$ and $s > 0$, we use the shorthand

$$\mathcal{V}(p, s) = \mathcal{V}_g(p, s) := \text{Vol}(B_g(p, s))$$

for the volume of the geodesic ball $B(p, s) = B_g(p, s)$ centred at p of radius s, computed with respect to the metric g. Our aim is to get a lower bound on the *volume ratio*

$$\mathcal{K}(p, r) := \frac{\mathcal{V}(p, r)}{r^n}, \qquad (8.3.1)$$

during a Ricci flow which will be turned into a lower bound for the injectivity radius, in terms of the maximum of the curvature $|\text{Rm}|$, in the next section.

Theorem 8.3.1 (No collapsing). *Suppose that $g(t)$ is a Ricci flow on a closed manifold $\mathcal{M}$, for $t \in [0, T]$. Working with respect to the metric $g(T)$, if $p \in \mathcal{M}$, and $r > 0$ is sufficiently small so that $|R| \le r^{-2}$ on $B(p, r)$, then*

$$\mathcal{K}(p, r) := \frac{\mathcal{V}(p, r)}{r^n} > \xi$$

for some $\xi > 0$ depending on n, $g(0)$, and upper bounds for r and T.

Remark 8.3.2.   We could replace the $|R| \le r^{-2}$ hypothesis by $|R| \le Cr^{-2}$, although ξ would then also depend on C.

Remark 8.3.3.   This is modelled on Perelman's "no collapsing" theorem from [31, §4] where a stronger hypothesis $|\mathrm{Rm}| \le r^{-2}$ is assumed. Further improvements are possible in which one assumes only bounds on a maximal function of the scalar curvature rather than pointwise curvature bounds – see [39] – although none of these improvements will be required in these notes.

We will prove the theorem above by combining an iteration argument with the theorem below.

Theorem 8.3.4.   *Suppose that $g(t)$ is a Ricci flow on a closed manifold $\mathcal{M}$, for $t \in [0, T]$, and that $r > 0$ and $p \in \mathcal{M}$. Then computing volumes, curvature and geodesic balls with respect to $g(T)$, we have*

$$\gamma \le \frac{\mathcal{V}(p, r)}{\mathcal{V}\left(p, \frac{r}{2}\right)} + \frac{r^2}{\mathcal{V}\left(p, \frac{r}{2}\right)} \int_{B(p,r)} |R| dV + \ln\left[\mathcal{K}(p, r)\right], \qquad (8.3.2)$$

for some $\gamma \in \mathbb{R}$ depending on n, $g(0)$, and upper bounds for r and T.

In the proof of this theorem, the monotonicity of $\mathcal{W}$ will give a lower bound for $\mathcal{W}$ under the flow, which will then be turned into geometric information via the following lemma.

Lemma 8.3.5.   *For any smooth metric g on a closed manifold $\mathcal{M}$, and $r > 0$, $p \in \mathcal{M}$ and $\lambda > 0$,*

$$\mu(g, \lambda r^2) \le 36\lambda \frac{\mathcal{V}(p, r) - \mathcal{V}\left(p, \frac{r}{2}\right)}{\mathcal{V}\left(p, \frac{r}{2}\right)} + \frac{\lambda r^2}{\mathcal{V}\left(p, \frac{r}{2}\right)} \int_{B(p,r)} |R| dV$$

$$+ \ln\left[\frac{\mathcal{V}(p, r)}{(4\pi\lambda r^2)^{\frac{n}{2}}}\right] - n.$$

Proof. (Lemma 8.3.5.)   If we adopt the change of variables we used in Section 8.1, namely $\phi = e^{-\frac{f}{2}}$, and write $\tau = \lambda r^2$, then we have (abusing notation for $\mathcal{W}$ as before) that

$$\mathcal{W}(g, \phi, \lambda r^2) = (4\pi\lambda r^2)^{-\frac{n}{2}} \int \left[\lambda r^2 (4|\nabla\phi|^2 + R\phi^2) - 2\phi^2 \ln\phi - n\phi^2\right] dV,$$

$$(8.3.3)$$

and the compatibility constraint (8.1.2) becomes

$$(4\pi\lambda r^2)^{-\frac{n}{2}} \int \phi^2 dV = 1. \qquad (8.3.4)$$

As we remarked before, these expressions have the benefit of making sense when ϕ is merely weakly (rather than strictly) positive, and by approximation,

$$\inf_f \mathcal{W}(g, f, \lambda r^2) = \inf_\phi \mathcal{W}(g, \phi, \lambda r^2), \qquad (8.3.5)$$

where the infima are taken over $f : \mathcal{M} \to \mathbb{R}$ and $\phi : \mathcal{M} \to [0, \infty)$ compatible with g and $\tau = \lambda r^2$, and we are abusing notation for $\mathcal{W}$ as usual.

Let $\psi : [0, \infty) \to [0, 1]$ be a smooth cut-off function, supported in $[0, 1]$, such that $\psi(y) = 1$ for $y \in [0, \frac{1}{2}]$ and $|\psi'| \leq 3$. We then write

$$\phi(x) = e^{-\frac{c}{2}} \psi \left(\frac{d(x, p)}{r} \right),$$

where $c \in \mathbb{R}$ is determined by the constraint (8.3.4), and since

$$\mathcal{V}\left(p, \frac{r}{2}\right) \leq e^c \int \phi^2 dV \leq \mathcal{V}(p, r),$$

we deduce that

$$(4\pi\lambda r^2)^{-\frac{n}{2}} \mathcal{V}\left(p, \frac{r}{2}\right) \leq e^c \leq (4\pi\lambda r^2)^{-\frac{n}{2}} \mathcal{V}(p, r). \qquad (8.3.6)$$

We now estimate each of the four terms in (8.3.3) separately.

Term 1. For the specific ϕ we have chosen, whose gradient is supported on $B(p, r) \backslash B(p, \frac{r}{2})$, and satisfies $|\nabla\phi| \leq e^{-\frac{c}{2}} \frac{1}{r} \sup |\psi'| \leq \frac{3}{r} e^{-\frac{c}{2}}$, we estimate

$$(4\pi\lambda r^2)^{-\frac{n}{2}} \int \lambda r^2 4|\nabla\phi|^2 dV \leq 4\lambda r^2 (4\pi\lambda r^2)^{-\frac{n}{2}} \sup |\nabla\phi|^2 \left(\mathcal{V}(p, r) - \mathcal{V}\left(p, \frac{r}{2}\right) \right)$$

$$\leq 36\lambda (4\pi\lambda r^2)^{-\frac{n}{2}} e^{-c} \left(\mathcal{V}(p, r) - \mathcal{V}\left(p, \frac{r}{2}\right) \right)$$

$$\leq 36\lambda \frac{\mathcal{V}(p, r) - \mathcal{V}\left(p, \frac{r}{2}\right)}{\mathcal{V}\left(p, \frac{r}{2}\right)},$$

the last inequality using the first part of (8.3.6).

Term 2. For the specific ϕ we have chosen, which is supported on $B(p, r)$, and satisfies

$$\phi^2 \leq e^{-c} \leq \frac{(4\pi\lambda r^2)^{\frac{n}{2}}}{\mathcal{V}\left(p, \frac{r}{2}\right)},$$

we estimate

$$(4\pi\lambda r^2)^{-\frac{n}{2}} \int \lambda r^2 R\phi^2 dV \leq \frac{\lambda r^2}{\mathcal{V}\left(p, \frac{r}{2}\right)} \int_{B(p,r)} |R| dV.$$

Term 3. By again using the fact that the support of ϕ lies within $B(p, r)$, we rewrite

$$(4\pi\lambda r^2)^{-\frac{n}{2}} \int -2\phi^2 \ln \phi \, dV = \int G(\sigma) d\mu, \qquad (8.3.7)$$

where $\sigma : B(p, r) \to [0, \infty)$ is defined by $\sigma := \phi^2$, the continuous function $G : [0, \infty) \to \mathbb{R}$ is defined for $y > 0$ by $G(y) := -y \ln y$, and the measure $d\mu$ is supported on $B(p, r)$, where $d\mu := (4\pi \lambda r^2)^{-\frac{n}{2}} dV$. Because G is concave, we may apply Jensen's inequality

$$\fint G(\sigma) d\mu \leq G\left(\fint \sigma d\mu\right),$$

as we used in Section 8.1, and since by (8.3.4) we have

$$\int \sigma d\mu = 1,$$

this tells us that

$$\int G(\sigma) d\mu \leq \left(\int d\mu\right) G\left(\frac{1}{\int d\mu}\right) = \ln\left(\int d\mu\right).$$

By (8.3.7) and the definition of $d\mu$, we conclude that

$$(4\pi \lambda r^2)^{-\frac{n}{2}} \int -2\phi^2 \ln \phi \, dV \leq \ln\left[\frac{\mathcal{V}(p, r)}{(4\pi \lambda r^2)^{\frac{n}{2}}}\right].$$

Term 4. By the constraint (8.3.4), we have simply

$$(4\pi \lambda r^2)^{-\frac{n}{2}} \int -n\phi^2 dV = -n.$$

When we combine these calculations with (8.3.3), we find that for the particular ϕ under consideration,

$$\mathcal{W}(g, \phi, \lambda r^2) \leq 36\lambda \frac{\mathcal{V}(p, r) - \mathcal{V}\left(p, \frac{r}{2}\right)}{\mathcal{V}\left(p, \frac{r}{2}\right)}$$
$$+ \frac{\lambda r^2}{\mathcal{V}\left(p, \frac{r}{2}\right)} \int_{B(p,r)} |R| dV + \ln\left[\frac{\mathcal{V}(p, r)}{(4\pi \lambda r^2)^{\frac{n}{2}}}\right] - n,$$

which together with (8.3.5), proves the lemma.   □

Proof. (Theorem 8.3.4.)   First, let us specialise Lemma 8.3.5 to the case $\lambda = \frac{1}{36}$ and $g = g(T)$, and estimate

$$\mu\left(g(T), \frac{1}{36} r^2\right)$$
$$\leq \frac{\mathcal{V}(p, r)}{\mathcal{V}\left(p, \frac{r}{2}\right)} + \frac{r^2}{\mathcal{V}\left(p, \frac{r}{2}\right)} \int_{B(p,r)} |R| dV + \ln [\mathcal{K}(p, r)] - \frac{n}{2} \ln \frac{\pi}{9} - n. \quad (8.3.8)$$

By Lemma 8.1.8, there exists a smooth $f_T : \mathcal{M} \to \mathbb{R}$ compatible with $g(T)$ and $\tau = \frac{1}{36}r^2$ such that

$$\mathcal{W}\left(g(T), f_T, \frac{1}{36}r^2\right) = \mu\left(g(T), \frac{1}{36}r^2\right). \tag{8.3.9}$$

We set $\tau = T + \frac{1}{36}r^2 - t$, and for our given Ricci flow $g(t)$, find the $f : \mathcal{M} \times [0, T] \to \mathbb{R}$ with $f(T) = f_T$ completing a solution of (8.2.1), as discussed in Remark 8.2.5. By Remark 8.2.2, $g(t)$, $f(t)$ and τ remain compatible for all t. Using the definition of μ, Proposition 8.2.1 and (8.3.9) we then have

$$\mu\left(g(0), \frac{1}{36}r^2 + T\right) \le \mathcal{W}\left(g(0), f(0), \frac{1}{36}r^2 + T\right) \le \mathcal{W}\left(g(T), f(T), \frac{1}{36}r^2\right)$$

$$= \mu\left(g(T), \frac{1}{36}r^2\right), \tag{8.3.10}$$

which coupled with (8.3.8) and the definition of ν from (8.1.7) gives

$$\nu\left(g(0), \frac{1}{36}r_0^2 + T_0\right) + \frac{n}{2}\ln\frac{\pi}{9} + n$$

$$\le \frac{\mathcal{V}(p,r)}{\mathcal{V}\left(p, \frac{r}{2}\right)} + \frac{r^2}{\mathcal{V}\left(p, \frac{r}{2}\right)}\int_{B(p,r)}|R|dV + \ln\left[\mathcal{K}(p,r)\right],$$

where r_0 and T_0 are any upper bounds for r and T respectively.   □

Proof. (Theorem 8.3.1.) Let r_0 and T_0 be upper bounds for r and T respectively. By Theorem 8.3.4, there exists $\gamma = \gamma(n, g(0), T_0, r_0) \in \mathbb{R}$ such that working with respect to the metric $g(T)$, we have, for all $s \in (0, r_0]$, that

$$\gamma \le \frac{\mathcal{V}(p,s)}{\mathcal{V}\left(p, \frac{s}{2}\right)} + \frac{s^2}{\mathcal{V}\left(p, \frac{s}{2}\right)}\int_{B(p,s)}|R|dV + \ln\left[\mathcal{K}(p,s)\right].$$

When $s \in (0, r]$, we have that $|R| \le r^{-2} \le s^{-2}$ on $B(p, s)$, and therefore

$$\gamma \le 2\frac{\mathcal{V}(p,s)}{\mathcal{V}\left(p, \frac{s}{2}\right)} + \ln\left[\mathcal{K}(p,s)\right]. \tag{8.3.11}$$

Define ω_n to be the volume of the unit ball in Euclidean n-space. Since $g(T)$ is a smooth metric, we have

$$\mathcal{K}(p, s) = \frac{\mathcal{V}(p, s)}{s^n} \to \omega_n \tag{8.3.12}$$

as $s \downarrow 0$. We claim that the theorem holds true if we choose

$$\xi := \min\left\{\frac{\omega_n}{2}, e^{\gamma - 2^{n+1}}\right\} > 0.$$

That is, we wish to establish that $\mathcal{K}(p, r) > \xi$.

Claim: If $s \in (0, r]$ and $\mathcal{K}(p, s) \leq \xi$ then $\mathcal{K}(p, \frac{s}{2}) \leq \xi$.

Indeed if $\mathcal{K}(p, s) \leq \xi$, then $\mathcal{K}(p, s) \leq e^{\gamma - 2^{n+1}}$, so $\ln[\mathcal{K}(p, s)] \leq \gamma - 2^{n+1}$, and by (8.3.11),

$$\gamma \leq 2 \frac{\mathcal{V}(p, s)}{\mathcal{V}(p, \frac{s}{2})} + \gamma - 2^{n+1}.$$

Rearranging, we find that

$$2^n \leq \frac{\mathcal{V}(p, s)}{\mathcal{V}(p, \frac{s}{2})} = 2^n \frac{\mathcal{K}(p, s)}{\mathcal{K}(p, \frac{s}{2})},$$

and hence that $\mathcal{K}(p, \frac{s}{2}) \leq \mathcal{K}(p, s) \leq \xi$, as required for the claim.

Using this claim iteratively, we see that

$$\mathcal{K}(p, r) \leq \xi \quad \Longrightarrow \quad \mathcal{K}(p, 2^{-m} r) \leq \xi \leq \frac{\omega_n}{2}$$

for all $m \geq 1$, but this contradicts the limit $\lim_{s \downarrow 0} \mathcal{K}(p, s) = \omega_n$ from (8.3.12).

$\square$

8.4 Volume ratio bounds imply injectivity radius bounds

The output of Theorem 8.3.1 ("no collapsing") is a positive lower bound on the volume ratio $\mathcal{K}$ at scales no larger than the inverse of the square-root of the curvature. In this section, we show that this implies a positive lower bound on the injectivity radius, as required to complete the discussion of singularity blow-ups that we began in Section 7.3.

Given a Riemannian manifold $(\mathcal{M}, g)$, we denote its injectivity radius by

$$\text{inj}(\mathcal{M}) = \text{inj}(\mathcal{M}, g) := \inf_{p \in \mathcal{M}} \text{inj}(\mathcal{M}, g, p).$$

Lemma 8.4.1. *There exist $\bar{r} > 0$ universal and $\eta > 0$ depending on the dimension n such that if $(\mathcal{M}^n, g)$ is a closed Riemannian manifold satisfying $|\text{Rm}| \leq 1$, then there exists $p \in \mathcal{M}$ such that*

$$\frac{\mathcal{V}(p, r)}{r^n} \leq \frac{\eta}{r} \text{inj}(\mathcal{M}), \qquad (8.4.1)$$

for all $r \in (0, \bar{r}]$.

In order to see the sharpness of this lemma, one can consider $\mathcal{M}$ to be the product of a closed manifold of dimension $n - 1$ satisfying $|\text{Rm}| \leq 1$, and a very small S^1.

Proof. For the moment, set $\bar{r} = 1$. We will arrive at the constant $\bar{r}$ whose existence is asserted in the proof by successively reducing it to satisfy a number of constraints.

First we observe that it suffices to prove the theorem under the assumption that $\text{inj}(\mathcal{M})$ is less than π. (We could take any positive upper bound here.) This follows because our curvature hypothesis $|\text{Rm}| \leq 1$ implies that all sectional curvatures are no lower than -1, and hence by Bishop's theorem (see, for example [14, (3.101)]) the volume ratio on the left-hand side of (8.4.1) is bounded above for $r \leq 1$ by some n-dependent constant C (which could be taken to be the volume ratio of the unit ball in hyperbolic space). Whenever $\text{inf}(\mathcal{M}) \geq \pi$, the theorem holds with $\eta = \frac{C}{\pi}$.

We now appeal to Klingenberg's lemma (see [4, Theorem 3.4]) which tells us that because all of the sectional curvatures are no more than 1, and $\text{inj}(\mathcal{M}) < \pi$, there exists a closed geodesic loop γ in $\mathcal{M}$ with

$$Length(\gamma) = 2\,\text{inj}(\mathcal{M}). \tag{8.4.2}$$

We are free to pick any $p \in \gamma$.

Let us define $N^r\gamma$ to be the subset of the normal bundle of γ within $\mathcal{M}$ consisting of vectors of length less than r, and $\mathcal{N}^r\gamma$ to be the subset of $\mathcal{M}$ consisting of all points within a distance r of γ. There is a natural flat metric G on the total space of $N^r\gamma$, which can be defined unambiguously at points over a neighbourhood $U \subset \gamma$ (small enough to be simply connected) of an arbitrary point $q \in \gamma$ as the pullback of the standard product metric on $\gamma \times N_q^r\gamma$ under the map

$$N^r U \rightarrow \gamma \times N_q^r U$$
$$v \rightarrow (\pi(v), \Theta(v)) \tag{8.4.3}$$

where $\pi : N^r\gamma \rightarrow \gamma$ is the bundle projection, and $\Theta(v)$ is the parallel translate of v within U to a vector over q.

We define a map $u : N^r\gamma \rightarrow \mathcal{N}^r\gamma$ by exponentiation. That is, for $v \in N^r\gamma$,

$$u(v) = \exp_{\pi(v)}(v).$$

Clearly, u is a surjection, since if $y \in \mathcal{N}^r\gamma$ and $q \in \gamma$ is a point closest to y, then y is in the image of an appropriate vector in the fibre over q. Moreover, by considering the behaviour of Jacobi fields under our constraint $|\text{Rm}| \leq 1$, and in particular by the Rauch comparison theorems (see [4]) after reducing the upper bound $\bar{r}$ for r, if necessary (to some positive, universal value) the map u is an immersion and we have

$$u^*g \leq 4G,$$

say. (Note that $u^*g = G$ on γ.) In particular, the volume form of u^*g is bounded by 2^n times the volume form of G, and hence

$$\text{Vol}_g(\mathcal{N}^r\gamma) \leq \text{Vol}_{u^*g}(N^r\gamma) \leq 2^n\,\text{Vol}_G(N^r\gamma) = 2^n Length(\gamma)\omega_{n-1}r^{n-1},$$

where ω_{n-1} is the volume of the $(n-1)$-dimensional Euclidean unit ball.

Clearly $B_g(p, r) \subset \mathcal{N}^r \gamma$, and so

$$\mathcal{V}(p, r) \leq \text{Vol}(\mathcal{N}^r \gamma) \leq 2^n \omega_{n-1} Length(\gamma) r^{n-1},$$

and by (8.4.2) we have

$$\frac{\mathcal{V}(p, r)}{r^n} \leq 2^{n+1} \omega_{n-1} \frac{\text{inj}(\mathcal{M})}{r}.$$

$\square$

8.5 Blowing up at singularities II

Now we have the "no collapsing" result of Theorem 8.3.1, and the injectivity radius control of Lemma 8.4.1, we can continue the discussion of blowing up Ricci flows near singularities that we began in Section 7.3.

Recall that for a Ricci flow $g(t)$ on a closed manifold $\mathcal{M}$, over a maximal time interval $[0, T)$ with $T < \infty$, we found appropriate sequences $\{p_i\} \subset \mathcal{M}$ and $t_i \uparrow T$, with

$$|\text{Rm}|(p_i, t_i) = \sup_{x \in \mathcal{M}, \, t \in [0, t_i]} |\text{Rm}|(x, t) \to \infty,$$

and defined blown-up Ricci flows

$$g_i(t) := |\text{Rm}|(p_i, t_i) g \left(t_i + \frac{t}{|\text{Rm}|(p_i, t_i)} \right)$$

such that, for all $a < 0$ and some $b = b(n) > 0$, the curvature of $g_i(t)$ was controlled for $t \in (a, b)$, and in particular,

$$\sup_{x \in \mathcal{M}, \, t \in (a,b)} |R(g_i(t))| \leq M < \infty,$$

for sufficiently large i (depending on a), and M dependent only on the dimension n. By choosing $r \in (0, M^{-\frac{1}{2}}]$, we have $|R(g_i(0))| \leq r^{-2}$. Converting back to the original flow, before blowing up, we see that if $0 < r \leq (M|\text{Rm}|(p_i, t_i))^{-\frac{1}{2}}$, then $|R(g(t_i))| \leq r^{-2}$.

By Theorem 8.3.1 we know that for all $p \in \mathcal{M}, 0 < r \leq (M|\text{Rm}|(p_i, t_i))^{-\frac{1}{2}}$ and sufficiently large i, we have the lower bound

$$\frac{\mathcal{V}(p, r)}{r^n} > \xi > 0$$

at time $t = t_i$, where $\xi = \xi(n, g(0), T) > 0$. Here, we included the condition "for sufficiently large i" so that we are free to ignore the dependency of ξ on an upper bound for r in Theorem 8.3.1.

Returning to the blown-up flows $g_i(t)$, we find that for all $p \in \mathcal{M}$, with respect to the metric $g_i(0)$,

$$\frac{\mathcal{V}(p, r)}{r^n} > \xi,$$

for all $r \in (0, M^{-\frac{1}{2}}]$.

By Lemma 8.4.1, after fixing $r = \min\{M^{-\frac{1}{2}}, \bar{r}\} > 0$ (a constant depending on n only) this implies the lower bound on the injectivity radius,

$$\mathrm{inj}(\mathcal{M}, g_i(0)) \geq \frac{r}{\eta} \frac{\mathcal{V}(p, r)}{r^n} > \frac{r\xi}{\eta} > 0.$$

In particular, $\frac{r\xi}{\eta}$ is a positive bound depending on n, $g(0)$ and T only. Therefore Theorem 7.2.3 allows us to pass to a subsequence to give the following conclusion.

Theorem 8.5.1 (Blow-up of singularities). *Suppose that $\mathcal{M}$ is a closed manifold, and $g(t)$ is a Ricci flow on a maximal time interval $[0, T)$ with $T < \infty$. Then there exist sequences $p_i \in \mathcal{M}$ and $t_i \uparrow T$ with*

$$|\mathrm{Rm}|(p_i, t_i) = \sup_{x \in \mathcal{M}, \, t \in [0, t_i]} |\mathrm{Rm}|(x, t) \to \infty,$$

such that, defining

$$g_i(t) := |\mathrm{Rm}|(p_i, t_i) \, g\left(t_i + \frac{t}{|\mathrm{Rm}|(p_i, t_i)}\right),$$

there exist $b = b(n) > 0$, a complete Ricci flow $(\mathcal{N}, \hat{g}(t))$ for $t \in (-\infty, b)$, and $p_\infty \in \mathcal{N}$ such that

$$(\mathcal{M}, g_i(t), p_i) \to (\mathcal{N}, \hat{g}(t), p_\infty)$$

as $i \to \infty$. Moreover, $|\mathrm{Rm}(\hat{g}(0))|(p_\infty) = 1$, and $|\mathrm{Rm}(\hat{g}(t))| \leq 1$ for $t \leq 0$.

9

Curvature pinching and preserved curvature properties under Ricci flow

9.1 Overview

We have already seen, in Corollary 3.2.3, that positive scalar curvature is preserved under the Ricci flow. Many other curvature conditions are preserved also, for example, positive curvature operator $\mathcal{R}$ [20].

We will focus on the three-dimensional case here, but we use techniques which extend to higher dimensions, albeit with some complications of notation. For the higher dimensional case, see [20]. We construct a general machinery for constraining the evolution of a tensor which obeys a nonlinear heat equation, and use it to prove that positive Ricci curvature is preserved, and that flows satisfying this condition must become very "round" wherever the curvature becomes large.

Later, in Theorem 10.2.1 we record another important application of these techniques, applicable to arbitrary flows in three dimensions, which constrains the blow-up Ricci flow $(\mathcal{N}, \hat{g}(t))$ of Theorem 8.5.1 to have weakly positive sectional curvature.

9.2 The Einstein Tensor, E

It is often useful to diagonalise the (symmetric) curvature operator $\mathcal{R}$: $\bigwedge^2 T\mathcal{M} \to \bigwedge^2 T\mathcal{M}$. In three dimensions, elements of $\bigwedge^2$ are all simple. (That is, we can write each of them as $e \wedge f$ for some vectors e and f, as opposed to in higher dimensions where, for example, $e_1 \wedge e_2 + e_3 \wedge e_4$ cannot be written in this way given linearly independent vectors $e_1, \ldots, e_4$.) It follows that, at each point $x \in \mathcal{M}$, $\mathcal{R}$ can be diagonalised by $e_2 \wedge e_3$, $e_3 \wedge e_1$, $e_1 \wedge e_2$ where e_1, e_2, e_3 form an orthonormal basis for $T_x\mathcal{M}$. All of the curvature information is then given by the eigenvalues of $\mathcal{R}$.

In fact, all the curvature information is given by the sectional curvatures λ_1, λ_2, λ_3 of the three "planes" $e_2 \wedge e_3$, $e_3 \wedge e_1$, $e_1 \wedge e_2$ respectively, because

$$\lambda_3 := \text{sectional curvature of } e_1 \wedge e_2 := \text{Rm}(e_1, e_2, e_1, e_2)$$

$$= \langle \mathcal{R}(e_1 \wedge e_2), e_1 \wedge e_2 \rangle$$

and thus these sectional curvatures are just the eigenvalues of $\mathcal{R}$ up to some constant[1].

An efficient way to handle this is in terms of the *Einstein tensor*, $E \in \Gamma(\text{Sym}^2 T^*\mathcal{M})$, defined by

$$E := -G(\text{Ric}) = -\text{Ric} + \frac{R}{2}g,$$

which we will normally view as a section $E \in \Gamma(T^*\mathcal{M} \otimes T\mathcal{M})$ using the metric. (That is, $E(X) = -\text{Ric}(X) + \frac{R}{2}X$, where we are keeping in mind Remark 2.3.8.) Note that Ric is diagonalised by $\{e_i\}$, with

$$\text{Ric}(e_i, e_j) = \begin{pmatrix} \lambda_2 + \lambda_3 & 0 & 0 \\ 0 & \lambda_1 + \lambda_3 & 0 \\ 0 & 0 & \lambda_1 + \lambda_2 \end{pmatrix}$$

and $R = 2(\lambda_1 + \lambda_2 + \lambda_3)$. Therefore E is also diagonalised by $\{e_i\}$, and

$$E(e_i, e_j) = \begin{pmatrix} \lambda_1 & 0 & 0 \\ 0 & \lambda_2 & 0 \\ 0 & 0 & \lambda_3 \end{pmatrix}.$$

Consequently, we may consider E instead of Rm.

9.3 Evolution of E under the Ricci flow

In Proposition 2.5.3, we recorded how the Ricci tensor evolves under Ricci flow. We may convert this to the equivalent formula for Ric seen as a section of $T^*\mathcal{M} \otimes T\mathcal{M}$ by computing

$$\left\langle \frac{\partial}{\partial t}\text{Ric}(X), W \right\rangle = \frac{\partial}{\partial t} \langle \text{Ric}(X), W \rangle + 2\text{Ric}(\text{Ric}(X), W)$$

$$= \frac{\partial}{\partial t}\text{Ric}(X, W) + 2 \langle \text{Ric}(X), \text{Ric}(W) \rangle$$

$$= \Delta\text{Ric}(X, W) + 2 \langle \text{Rm}(X, \cdot, W, \cdot), \text{Ric} \rangle$$

---

[1] This constant depends on conventions which we are otherwise able to leave ambiguous. First, it depends on whether you use the usual inner product on $\bigwedge^2$ which makes $\langle e_1 \wedge e_2, e_1 \wedge e_2 \rangle = 1$, or keep using the usual tensor inner product which would make $\langle e_1 \otimes e_2, e_1 \otimes e_2 \rangle = 1$. In the latter case, one would have to decide on a convention for the wedge product – that is, whether $e_1 \wedge e_2 = e_1 \otimes e_2 - e_2 \otimes e_1$ or $e_1 \wedge e_2 = \frac{1}{2}(e_1 \otimes e_2 - e_2 \otimes e_1)$.

Moreover $\frac{\partial R}{\partial t} = \Delta R + 2|\mathrm{Ric}|^2$ and thus

$$
\begin{aligned}
\left\langle \frac{\partial E}{\partial t}(X), W \right\rangle &= -\Delta\mathrm{Ric}(X, W) - 2\langle \mathrm{Rm}(X, \cdot, W, \cdot), \mathrm{Ric}\rangle \\
&\quad + \frac{1}{2}(\Delta R)g(X, W) + |\mathrm{Ric}|^2 g(X, W) \\
&= \Delta E(X, W) - 2\langle \mathrm{Rm}(X, \cdot, W, \cdot), \mathrm{Ric}\rangle + |\mathrm{Ric}|^2 g(X, W).
\end{aligned}
$$
$$(9.3.1)$$

Writing this with respect to the basis $\{e_i\}$ then gives

$$
\frac{\partial E}{\partial t} - \Delta E = 2 \begin{pmatrix} \lambda_1^2 + \lambda_2\lambda_3 & 0 & 0 \\ 0 & \lambda_2^2 + \lambda_1\lambda_3 & 0 \\ 0 & 0 & \lambda_3^2 + \lambda_1\lambda_2 \end{pmatrix}.
\qquad (9.3.2)
$$

There are a number of ways of writing the right-hand side explicitly in terms of E. For example, it represents the fibre-gradient of the function $\frac{2}{3}\mathrm{tr}\,(E^3) + 2\det(E)$ on $T^*\mathcal{M} \otimes T\mathcal{M}$, or can be written in terms of the Lie-algebra square of E (see [20] for the latter viewpoint) but we are happy to leave it in this form for now.

9.4 The Uhlenbeck Trick

We have arrived at a relatively simple evolution equation for E, and hence for the curvature. We wish to apply a maximum principle to find curvature conditions which are preserved under the Ricci flow. One remaining obstacle is that it will be important to see E not just as a section of $T^*\mathcal{M} \otimes T\mathcal{M}$, but as a section of the sub-bundle of such sections which are symmetric with respect to $g(t)$. (That is, $g(E(X), Y) = g(E(Y), X)$ for all vector fields X and Y.) Unfortunately, this sub-bundle inherits the t-dependence of $g(t)$, which would cause problems. In this section, we use the so-called 'Uhlenbeck trick' to hide this t-dependence.

Suppose we have a Riemannian manifold $(\mathcal{M}, g_0)$. Let V be an abstract vector bundle over $\mathcal{M}$, isomorphic to $T\mathcal{M}$ via a bundle isomorphism $u_0 : V \to T\mathcal{M}$, and with fibre metric $h = u_0^*(g_0)$ (that is, $h(v, w) = g_0(u_0(v), u_0(w))$ for all sections $v, w \in \Gamma(V)$) so that u_0 is a bundle isometry.

Remark 9.4.1. We could just write $T\mathcal{M}$ in place of V, but then we might be tempted to use, say, the Levi-Civita connection of a Riemannian metric on $\mathcal{M}$, and we don't want to allow such extra structure to be used.

Let $g(t)$ be the Ricci flow on $\mathcal{M}$ with $g(0) = g_0$ over some time interval $[0, T]$. Consider, for $t \in [0, T]$, the one-parameter family of bundle endomorphisms

$u = u_t : V \to T\mathcal{M}$ solving

$$\begin{cases} \frac{\partial u}{\partial t} = \text{Ric}_{g(t)}(u) \\ u(0) = u_0. \end{cases} \qquad (9.4.1)$$

By (9.4.1) we mean that if $v \in \Gamma(V)$ then $u(v)$ evolves under the equation $\frac{\partial}{\partial t}u(v) = \text{Ric}_{g(t)}(u(v))$. For all sections $v, w \in \Gamma(V)$, we then have

$$\frac{\partial}{\partial t}g(u(v), u(w)) = -2\text{Ric}(u(v), u(w)) + g\left(\frac{\partial}{\partial t}u(v), u(w)\right)$$

$$+ g\left(u(v), \frac{\partial}{\partial t}u(w)\right)$$

$$= -2\text{Ric}(u(v), u(w)) + g\left(\text{Ric}(u(v)), u(w)\right)$$

$$+ g\left(u(v), \text{Ric}(u(w))\right)$$

$$= -2\text{Ric}(u(v), u(w)) + 2\text{Ric}(u(v), u(w))$$

$$= 0.$$

Therefore, for all t,

$$g(u(v), u(w)) = g_0(u_0(v), u_0(w)) = h(v, w),$$

and thus $h = u_t^*(g(t))$ for all t, showing that u remains a bundle isometry $(V, h) \mapsto (T\mathcal{M}, g(t))$. The metric h extends as usual to tensor products of V and V^*.

Now define a connection $A(t)$ on V to be the pull-back at each time of the Levi-Civita connection on $T\mathcal{M}$ under u. That is,

$$u(A_X v) = \nabla_X(u(v))$$

for all $X \in \Gamma(T\mathcal{M})$ and $v \in \Gamma(V)$. The connection A extends as usual (via the product rule etc.) to a connection on tensor products of V and V^*. For example

$$A(v \otimes w) = (Av) \otimes w + v \otimes (Aw), \quad \text{and} \quad X(\theta(v)) = (A_X \theta)v + \theta(A_X v)$$

for $v, w \in \Gamma(V)$, $\theta \in \Gamma(V^*)$ and $X \in \Gamma(T\mathcal{M})$, where we are suppressing the obvious juggling of the order of the tensor entries in the first expression. Moreover, by combining A with the Levi-Civita connection on $T\mathcal{M}$, we can extend $A(t)$ to a connection, also denoted by $A(t)$, on tensor products of V, V^*, $T\mathcal{M}$ and $T^*\mathcal{M}$. For example, $A(v \otimes X) = (Av) \otimes X + v \otimes \nabla X$ for $v \in \Gamma(V)$ and $X \in \Gamma(T\mathcal{M})$. In particular, it makes sense to consider the Laplacian $\Delta_A = \text{tr}A^2$ on such products.

Remark 9.4.2. By considering u as an element of $\Gamma(V^* \otimes T\mathcal{M})$, we can apply A to u and find that $Au = 0$. Indeed, because

$$u(A_X v) = \nabla_X(u(v)) = A_X(u(v)) = (A_X u)v + u(A_X v),$$

it follows that $A_X u = 0$ for all $X \in \Gamma(T\mathcal{M})$.

Remark 9.4.3. Seeing h as an element of $\Gamma(V^* \otimes V^*)$, we also have $Ah = 0$ — that is, A is compatible with h – by construction, or by calculating

$$
\begin{aligned}
(A_X h)(v, w) &= X(h(v, w)) - h(A_X v, w) - h(v, A_X w) \\
&= X[g(u(v), u(w))] - g(u(A_X v), u(w)) - g(u(v), u(A_X w)) \\
&= g(\nabla_X(u(v)), u(w)) + g(u(v), \nabla_X(u(w))) \\
&\quad - g(\nabla_X(u(v)), u(w)) - g(u(v), \nabla_X(u(w))) \\
&= 0,
\end{aligned}
$$

for all $X \in \Gamma(T\mathcal{M})$, $v, w \in \Gamma(V)$.

Let us check how the Einstein tensor $E \in \Gamma(T^*\mathcal{M} \otimes T\mathcal{M})$ looks when we pull it back under u. In other words, we consider the section $\tilde{E} \in \Gamma(V^* \otimes V)$ defined by

$$u(\tilde{E}(v)) = E(u(v))$$

for $v \in \Gamma(V)$. The notable advantage of $\tilde{E}$ over E is that it can be seen as a section of the sub-bundle of endomorphisms which are *symmetric* with respect to h rather than $g(t)$. Since h is independent of t, this sub-bundle is also independent of t.

The eigenvalues of $\tilde{E}$, which can be seen as three functions on $\mathcal{M}$, are precisely the eigenvalues of E (and so we still call them λ_1, λ_2 and λ_3) with eigenvectors identified via the bundle isomorphism u.

Moreover, for all $v \in \Gamma(V)$, we have

$$u(\Delta_A \tilde{E}(v)) = \Delta E(u(v))$$

since $Au \equiv 0$, and

$$
\begin{aligned}
u\left(\frac{\partial \tilde{E}}{\partial t}(v) \right) &= \frac{\partial}{\partial t}[u(\tilde{E}(v))] - \frac{\partial u}{\partial t}(\tilde{E}(v)) \\
&= \frac{\partial}{\partial t}[E(u(v))] - \operatorname{Ric}(u(\tilde{E}(v))) \\
&= \frac{\partial E}{\partial t}(u(v)) + E(\frac{\partial u}{\partial t}(v)) - \operatorname{Ric}(E(u(v))) \\
&= \frac{\partial E}{\partial t}(u(v)) + E(\operatorname{Ric}(u(v))) - \operatorname{Ric}(E(u(v))) \\
&= \frac{\partial E}{\partial t}(u(v))
\end{aligned}
$$

where the last equality follows because Ric and E commute. Combining, we may write, with respect to the eigenvectors of $\tilde{E}$,

$$\frac{\partial \tilde{E}}{\partial t} - \Delta_A \tilde{E} = 2 \begin{pmatrix} \lambda_1^2 + \lambda_2 \lambda_3 & 0 & 0 \\ 0 & \lambda_2^2 + \lambda_1 \lambda_3 & 0 \\ 0 & 0 & \lambda_3^2 + \lambda_1 \lambda_2 \end{pmatrix}, \qquad (9.4.2)$$

which is the same equation as before, but with the advantage, as described earlier that the fibre metric h is independent of time, and so the bundle of symmetric endomorphisms is now also independent of time. From now on, we will drop the $\sim$ over the E.

9.5 Formulae for parallel functions on vector bundles

In the general situation that E is a section of a vector bundle satisfying a nonlinear heat equation, we wish to develop a method for constraining E to remain in certain subsets of the bundle, using only the scalar maximum principle from Chapter 3. We will lay the groundwork in this section, and prove a so-called ODE-PDE theorem in the next.

Let W be a vector bundle over a manifold $\mathcal{M}$, the former equipped with a fixed fibre metric h, and the latter equipped with a time-dependent Riemannian metric $g(t)$, for $t \in [0, T]$. Let $A(t)$ be a time-dependent connection on this bundle, for $t \in [0, T]$, which is compatible at each time with the metric h. We may extend this connection to sections of tensor products of W, W^*, $T\mathcal{M}$ and $T^*\mathcal{M}$ in the usual way.

Remark 9.5.1. The only application of this construction in these notes will be to the case where $g(t)$ is a Ricci flow on a closed manifold $\mathcal{M}$, and, in terms of the bundle V of the last section, with W the sub-bundle of $V^* \otimes V$ consisting of endomorphisms which are *symmetric* with respect to the metric h of the last section. This metric h and its compatible connection $A(t)$ may be extended to W to give the metric and connection required in this section.

Definition 9.5.2. A (smooth) function $\Psi : W \to \mathbb{R}$ is said to be *parallel* if it is invariant under parallel translation using the connection $A(t)$, at each time $t \in [0, T]$. That is, if $w_1 \in W$ can be parallel-translated into $w_2 \in W$ (which may have a different base-point) at some time $t \in [0, T]$, then $\Psi(w_1) = \Psi(w_2)$.

Now suppose that $E(t) \in \Gamma(W)$ is a (smooth) time-dependent section, for $t \in [0, T]$, and consider the function $\Psi \circ E : \mathcal{M} \times [0, T] \to \mathbb{R}$ for some parallel function Ψ. We wish to express the evolution of this function in terms of the evolution of the section E.

Let us fix $t \in [0, T]$ for the moment, dropping t as an argument in expressions. If we fix a frame $\mathfrak{e}_1, \ldots, \mathfrak{e}_l$ in W_p, and extend it to a smooth time-dependent local frame for W by radial parallel translation using the connection A, then we can check by working in normal coordinates that at p, we have

$$A\mathfrak{e}_\alpha = 0; \qquad A^2\mathfrak{e}_\alpha = -\frac{1}{2}R_A(\cdot, \cdot)\mathfrak{e}_\alpha, \qquad (9.5.1)$$

for each $\alpha \in \{1, \ldots, l\}$, where R_A is the curvature of the connection A. The section $E \in \Gamma(W)$ may locally be written

$$E(x) = \sum_{\alpha=1}^{l} a^\alpha(x)\mathfrak{e}_\alpha(x),$$

for x in a neighbourhood of p, and then by (9.5.1), we have that

$$A^2 E(p) = \left(\sum_{\alpha=1}^{l} \nabla da^\alpha(p)\mathfrak{e}_\alpha(p)\right) - \frac{1}{2}R_A(\cdot, \cdot)E(p). \qquad (9.5.2)$$

It will be convenient to use, for each $p \in \mathcal{M}$, the notation $\Psi_p : W_p \to \mathbb{R}$ for the restriction of Ψ to the fibre W_p at p. Using the fact that Ψ is parallel, we have

$$\Psi(E(x)) = \Psi\left(\sum_{\alpha=1}^{l} a^\alpha(x)\mathfrak{e}_\alpha(x)\right) = \Psi_p\left(\sum_{\alpha=1}^{l} a^\alpha(x)\mathfrak{e}_\alpha(p)\right), \qquad (9.5.3)$$

for x near p. We can use this expression to compute the Laplacian of $\Psi \circ E :$ $\mathcal{M} \to \mathbb{R}$. First, we have

$$d(\Psi \circ E)(x) = d\left(\Psi_p\left(\sum_{\alpha=1}^{l} a^\alpha(x)\mathfrak{e}_\alpha(p)\right)\right)$$
$$= d\Psi_p\left(\sum_{\alpha=1}^{l} a^\alpha(x)\mathfrak{e}_\alpha(p)\right)\left(\sum_{\alpha=1}^{l} da^\alpha(x)\mathfrak{e}_\alpha(p)\right). \qquad (9.5.4)$$

One could evaluate this at $x = p$ to give $d(\Psi \circ E)(p) = d\Psi_p(E(p))(AE(p))$. Instead, we take a further derivative, and evaluate at p, to give

$$\nabla d(\Psi \circ E)(p) = \text{Hess}(\Psi_p)(E(p))(AE(p), AE(p))$$
$$+ d\Psi_p(E(p))\left(A^2E(p) + \frac{1}{2}R_A(\cdot, \cdot)E(p)\right), \qquad (9.5.5)$$

by (9.5.2). Note that the function Ψ_p is defined on the *vector space* W_p, so its Hessian $\text{Hess}(\Psi_p)$ is defined in the classical sense. If we assume that Ψ_p is convex (by which we always mean weakly convex) then after taking the trace of (9.5.5) one sees that

$$\Delta_{\mathcal{M}}(\Psi \circ E)(p) \geq d\Psi_p(E(p))(\Delta_A E(p)). \qquad (9.5.6)$$

The convexity of Ψ_p implies the convexity of Ψ_q for every other $q \in \mathcal{M}$ because Ψ is parallel, allowing us to make the following definition.

Definition 9.5.3. A function $\Psi : W \to \mathbb{R}$ which is parallel in the sense of Definition 9.5.2 will be called *convex* if Ψ_p is a convex function for some $p \in \mathcal{M}$ or equivalently for all $p \in \mathcal{M}$.

Reinstating the t-dependence of E, we may also calculate

$$\frac{\partial}{\partial t}(\Psi \circ E)(p, t) = \frac{\partial}{\partial t}\Psi_p(E(p, t)) = d\Psi_p(E(p, t))\left(\frac{\partial E}{\partial t}(p, t)\right), \quad (9.5.7)$$

and combining this with (9.5.6) gives the following result.

Proposition 9.5.4. *Suppose that W is a vector bundle over $(\mathcal{M}, g(t))$ with connection $A(t)$, and $E(t) \in \Gamma(W)$ is some section, with smooth dependence on $t \in [0, T]$. Then for any function $\Psi : W \to \mathbb{R}$ which is parallel in the sense of Definition 9.5.2 and convex in the sense of Definition 9.5.3, we have*

$$\left(\frac{\partial}{\partial t} - \Delta_\mathcal{M}\right)(\Psi \circ E)(x, t) \leq d\Psi_x(E(x, t))\left(\frac{\partial E}{\partial t} - \Delta_A E\right)(x, t) \quad (9.5.8)$$

Finally, we extend the definitions of *parallel* and *convex* from functions on W to subsets of W.

Definition 9.5.5. A subset $X \subset W$ is said to be parallel if it is invariant under parallel translation using the connection $A(t)$ at each time $t \in [0, T]$. That is, whenever $w_1 \in X$ can be parallel translated to a point $w_2 \in W$ (which may have a different base point) then $w_2 \in X$. Such a subset is said to be convex if its intersection with one fibre (equivalently all fibres) of W is convex.

In practice, all such sets will arise as sublevel sets of parallel convex functions on W, or intersections thereof.

9.6 An ODE-PDE theorem

By applying the maximum principle to the composition of appropriate smooth, parallel, convex functions Ψ with evolving sections E, we will be able to constrain the evolution of E to remain within sublevel sets of Ψ, or more generally, within arbitrary convex subsets of W, when E satisfies an appropriate PDE. This result comes from [20], where the proof is presented in a somewhat different fashion.

Theorem 9.6.1. *Suppose that W is a vector bundle over a closed manifold $\mathcal{M}$, with bundle projection $\pi : W \to \mathcal{M}$, equipped with a fixed fibre metric h and* a smooth family of connections $A(t)$ compatible with h, for $t \in [0, T]$. Suppose that $g(t)$ is a smooth family of Riemannian metrics on $\mathcal{M}$, for $t \in [0, T]$.

Suppose that $\Upsilon \in \Gamma(\pi^*(W))$, that is, Υ is a vector field on W which is parallel to the fibres of W, and that $X \subset W$ is a closed subset which is parallel and convex in the sense of Definition 9.5.5, and preserved under the ODE

$$\frac{d\,e}{dt} = \Upsilon(e). \tag{9.6.1}$$

More precisely, if $e(t)$ is a solution to (9.6.1) on some time interval $[0, \epsilon]$, with $e(0) \in X$ then we assume that $e(t) \in X$ for all t.

Then the set X is preserved under the PDE

$$\frac{\partial E}{\partial t} - \Delta_A E = \Upsilon(E), \tag{9.6.2}$$

for $E(t) \in \Gamma(W)$, $t \in [0, T]$. In other words, if $E(x, 0) \in X$ for all $x \in \mathcal{M}$, then $E(x, t) \in X$ for all $x \in \mathcal{M}$ and $t \in [0, T]$.

Note that the ODE (9.6.1) preserves each fibre.

Proof. First note that we may assume, without loss of generality, that X is compact. Indeed, if we had a counterexample to the theorem for X noncompact, then we could intersect X with the parallel, convex subset

$$T_r := \{w \in W \; : \; |w| < r\}, \tag{9.6.3}$$

for $r > 0$ sufficiently large that the image of the counterexample E always lay within $T_{r/2}$, and modify Υ outside the image of E appropriately to give a counterexample for X compact. (Here, and throughout the proof, the quantity $|w|$ of a point in W will always be measured with respect to h.)

Define $\psi : W \to \mathbb{R}$ by $\psi(w) = dist(w, X)$, the fibrewise distance function to the convex set X, using the fibre metric h. Define $Y = \{w \in W \mid \psi(w) < 2\}$, and let K be the (fibrewise) Lipschitz constant for Υ within Y. That is, for all $x \in \mathcal{M}$, and for all $v, w \in Y_x$ (the part of Y in the fibre over x) we have

$$|\Upsilon(v) - \Upsilon(w)| \le K|v - w|. \tag{9.6.4}$$

For a given $w \in Y \backslash X$, we pick the point $p \in \partial X$ in the same fibre as w which is closest to w, and define $x := \pi(w)$. It is also convenient to highlight the linear function $L : W_x \to \mathbb{R}$ defined by $L(v) = d\psi_x(w)(v - p)$, and note that

$$L \le 0 \text{ on } X_x; \qquad L \le \psi \le dist(\cdot, p) \text{ near } w \text{ in } W_x. \tag{9.6.5}$$

By the ODE hypothesis, we must have

$$d\psi_x(w)(\Upsilon(p)) \le 0. \tag{9.6.6}$$

If not, we would solve the ODE (9.6.1) with $e(0) = p$ to find that

$$\frac{d}{dt} L(e(t)) \bigg|_{t=0} = d\psi_x(w)(\Upsilon(p)) > 0,$$

which, by (9.6.5), would make $\psi(e(t)) \geq L(e(t)) > 0$ for small $t > 0$, a contradiction.

The inequality (9.6.6) then leads to

$$\begin{aligned} d\psi_{\pi(w)}(w)(\Upsilon(w)) &= d\psi_x(w)(\Upsilon(w) - \Upsilon(p)) + d\psi_x(w)(\Upsilon(p)) \\ &\leq |\Upsilon(w) - \Upsilon(p)| \leq K\psi(w), \end{aligned} \tag{9.6.7}$$

for our arbitrary point $w \in Y \backslash X$.

In order to apply the maximum principle as we stated it in Chapter 3, it will be convenient to work with a slightly smoothed version of ψ. Since ψ is C^1 on $Y \backslash X$ (recall (9.6.5)) for any $\varepsilon \in (0, 1)$, we can mollify ψ (see [15]) by a small amount (depending on ε) on each fibre, using the metric h, to give a smooth, parallel, convex function $\Psi : W \to \mathbb{R}$ for which

$$|\Psi - \psi| < \varepsilon$$

on Y, and

$$d\Psi_{\pi(w)}(w)(\Upsilon(w)) \leq 2K\Psi(w) \tag{9.6.8}$$

on the set $\{w \in W \mid \Psi(w) \in [\frac{\varepsilon}{2}, 1]\} \subset Y$.

Now let us take a solution E to the PDE as in the theorem. We will constrain the evolution of E by applying the maximum principle to $u := \Psi \circ E : \mathcal{M} \times [0, T] \to \mathbb{R}$. By Proposition 9.5.4 and (9.6.8), we have

$$\left(\frac{\partial}{\partial t} - \Delta_{\mathcal{M}} \right) u(x, t) \leq d\Psi_x(E(x, t))(\Upsilon(E(x, t))) \leq 2Ku(x, t) \tag{9.6.9}$$

whenever $u(x, t) \in [\frac{\varepsilon}{2}, 1]$. Moreover, if we choose some smooth $F : \mathbb{R} \to \mathbb{R}$ for which $F(v) = 2Kv$ for $v \geq \varepsilon$, and so that $d\Psi_{\pi(w)}(w)(\Upsilon(w)) \leq F(\Psi(w))$ for all $w \in W$ with $\Psi(w) \leq 1$, then

$$\left(\frac{\partial}{\partial t} - \Delta_{\mathcal{M}} \right) u \leq F(u) \tag{9.6.10}$$

for sufficiently small t so that $u(\cdot, t) \leq 1$. This is the equation to which we can apply Theorem 3.1.1. Noting that $u(\cdot, 0) \leq \varepsilon$, we compare u to the function $\phi(t) = \varepsilon e^{2Kt}$, and find that for $\varepsilon < e^{-2KT}$, we have $u(\cdot, t) \leq \varepsilon e^{2Kt} \leq \varepsilon e^{2KT} < 1$, the last of these inequalities being mentioned to emphasise that we need not worry about (9.6.10) failing to hold before time T. Returning from Ψ to ψ, we find that $\psi \circ E \leq \Psi \circ E + \varepsilon = u + \varepsilon$, and hence that

$$\psi \circ E \leq \varepsilon(e^{2KT} + 1)$$

on $\mathcal{M} \times [0, T]$. This last inequality is independent of the mollification, and holds for all sufficiently small $\varepsilon > 0$, and hence $\psi \circ E \equiv 0$ throughout, which is equivalent to having $E(x, t) \in X$ for all $x \in \mathcal{M}$ and $t \in [0, T]$ as desired.

$\square$

The only aspect of the ODE hypothesis that we used in the proof is the following property: *For all $p \in \partial X$, when we solve the ODE (9.6.1) with $e(0) = p$,* then

$$\frac{d}{dt}L(e(t))\bigg|_{t=0} \leq 0 \qquad (9.6.11)$$

whenever $L : W_x \to \mathbb{R}$ is a nonconstant linear function such that $L(p) = 0$ and $L \leq 0$ on X. This property could be shown to be equivalent to the ODE hypothesis, although one should beware of the subtlety that if we only knew that for *some* $p \in \partial X$, the ODE solution $e(t)$ with $e(0) = p$ satisfied (9.6.11) for all L as before, then it could still be the case that $e(t) \notin X$ for all $t > 0$ even though $e(0) = p \in X$.

The following consequence of these considerations will be useful:

Remark 9.6.2.   If X is defined as the 0-sublevel set $\Psi^{-1}((-\infty, 0])$ of a convex, parallel function $\Psi : W \to \mathbb{R}$ whose minimum is negative, then the ODE hypothesis is satisfied if for all $w \in W$ with $\Psi(w) = 0$, the solution of the ODE (9.6.1) with $e(0) = w$ satisfies

$$\frac{d}{dt}\Psi(e(t))\bigg|_{t=0} \leq 0.$$

9.7 Applications of the ODE-PDE theorem

Remark 9.7.1.   We will apply Theorem 9.6.1 in the case described in Remark 9.5.1. We take the section Υ so that the PDE (9.6.2) for E is the PDE (9.4.2) we had for $\tilde{E}$. Note that in this situation, the ODE (9.6.1) preserves the eigenvectors of the symmetric endomorphism e, and thus can be viewed simply as the coupled ODE

$$\frac{d\lambda_1}{dt} = 2(\lambda_1^2 + \lambda_2\lambda_3), \quad \frac{d\lambda_2}{dt} = 2(\lambda_2^2 + \lambda_3\lambda_1), \quad \frac{d\lambda_3}{dt} = 2(\lambda_3^2 + \lambda_1\lambda_2)$$

$$(9.7.1)$$

for the eigenvalues λ_1, λ_2 and λ_3.

By definition of the extension of $A(t)$ to W from V, and the fact that $A(t)$ is compatible with the metric h, when we parallel translate an element of W (that is, an endomorphism) the eigenvectors in V move by parallel translation, and the eigenvalues remain the same.

Consequently, if we define a function Ψ on W exclusively in terms of the eigenvalues $\lambda_1, \lambda_2, \lambda_3$, it will automatically be a parallel function, in the sense of Definition 9.5.2.

In order to make Ψ convex, in the sense of Definition 9.5.3, it suffices to choose a function of $\lambda_1, \lambda_2, \lambda_3$ which gives a convex function on the space of symmetric endomorphisms of $\mathbb{R}^3$ endowed with the standard metric. This motivates the following lemma.

Lemma 9.7.2 (Convexity). *Considering eigenvalues $\lambda_1 \leq \lambda_2 \leq \lambda_3$ as functions on the vector space of symmetric endomorphisms of $\mathbb{R}^3$, we have*

(i) $\lambda_1 + \lambda_2 + \lambda_3$ is linear (convex and concave);

(ii) λ_1 is concave.

As a consequence, we have that

(a) λ_3 is convex;

(b) $\lambda_1 + \lambda_2$ is concave;

(c) $\lambda_3 - \lambda_1$ is convex.

Moreover, if $\alpha : \mathcal{X} \to \mathbb{R}$ is a concave function on a vector space $\mathcal{X}$, and $\beta : [0, \infty) \to \mathbb{R}$ is concave and increasing, then $\beta \circ \alpha : \alpha^{-1}([0, \infty)) \subset \mathcal{X} \to \mathbb{R}$ is concave.

Proof.

(i) $\lambda_1 + \lambda_2 + \lambda_3$ is just the *trace* of the symmetric endomorphism, which is a linear function.

(ii) Suppose that F_1 and F_2 are two symmetric endomorphisms on $\mathbb{R}^3$, and that $s \in [0, 1]$. Let $e \in \mathbb{R}^3$ be a unit-length eigenvector corresponding to the first eigenvalue of $sF_1 + (1 - s)F_2$. Then

$$\lambda_1(sF_1 + (1 - s)F_2) = s \langle F_1(e), e \rangle + (1 - s) \langle F_2(e), e \rangle$$
$$\geq s\lambda_1(F_1) + (1 - s)\lambda_1(F_2)$$

since $\lambda_1(F_1) = \inf_{\|\tilde{e}\|=1} \langle F_1(\tilde{e}), \tilde{e} \rangle$.

(a) This follows from (ii) since if E is a symmetric endomorphism on $\mathbb{R}^3$, then $\lambda_3(E) = -\lambda_1(-E)$.

(b) This follows from (i) and (a) because $\lambda_1 + \lambda_2 = (\lambda_1 + \lambda_2 + \lambda_3) - \lambda_3$.

(c) This follows from (ii) and (a).

For the final claim, since the function $\beta : [0, \infty) \to \mathbb{R}$ is concave, for $x_1, x_2 \in \alpha^{-1}([0, \infty)) \subset \mathcal{X}$ we have

$$s\beta(\alpha(x_1)) + (1 - s)\beta(\alpha(x_2)) \leq \beta[s\alpha(x_1) + (1 - s)\alpha(x_2)]$$
$$\leq \beta[\alpha(sx_1 + (1 - s)x_2)],$$

where the final inequality uses the monotonicity of β. $\qquad\square$

We will use this lemma to prove a variety of curvature pinching results. The important result for the next chapter is Theorem 9.7.8, whose proof relies on Theorem 9.7.5.

Theorem 9.7.3.   *Weakly positive sectional curvature is preserved under Ricci flow on three-dimensional closed manifolds.*

Remark 9.7.4.   A minor adaptation of the proof would show that strictly positive sectional curvature is also preserved in this case.

Proof.   We apply Theorem 9.6.1 in the situation described in Remark 9.7.1, with X the 0-sublevel set of the convex function $\Psi(e) = -\lambda_1(e)$. (That is, $X = \{e \in W \mid \Psi(e) \leq 0\}$.) The theorem tells us that the condition $\lambda_1 \geq 0$ is preserved under the Ricci flow (and hence that weakly positive sectional curvature is preserved, as desired) provided that the same condition is preserved under the analogous ODE (9.6.1), or equivalently (9.7.1). The λ_1 component of this ODE is simply

$$\frac{d\lambda_1}{dt} = 2(\lambda_1^2 + \lambda_2\lambda_3),$$

and thus keeping in mind that $\lambda_2, \lambda_3 \geq \lambda_1 \geq 0$, we see that $\lambda_1 \geq 0$ is preserved.
$\square$

In addition to this, positive *Ricci* curvature is also preserved in three dimensions. We require a slightly sharper fact.

Theorem 9.7.5.   *For all $\varepsilon \in [0, \frac{1}{3})$ and any Ricci flow $g(t)$ on a three-dimensional, closed manifold $\mathcal{M}$, the "pinching condition"*

$$\mathrm{Ric} \geq \varepsilon R g \qquad\qquad (9.7.2)$$

is preserved.

Refer back to Remark 5.3.3 for the definition of such inequalities, if necessary.

The significance of the permitted range of values for ε is that (9.7.2) then implies $R \geq 0$, and hence that $\mathrm{Ric} \geq 0$, by tracing.

Proof.   The eigenvalues of Ric are (in increasing order)

$$\lambda_1 + \lambda_2, \quad \lambda_1 + \lambda_3, \quad \lambda_2 + \lambda_3,$$

and in particular, $R = 2(\lambda_1 + \lambda_2 + \lambda_3)$. Define $\delta := \frac{2\varepsilon}{1-2\varepsilon} \in [0, 2)$. Therefore

$$\mathrm{Ric} \geq \varepsilon R g \Leftrightarrow \lambda_1 + \lambda_2 \geq \varepsilon(2(\lambda_1 + \lambda_2 + \lambda_3))$$

$$\Leftrightarrow \lambda_1 + \lambda_2 \geq \frac{2\varepsilon}{1 - 2\varepsilon}\lambda_3$$

$$\Leftrightarrow \delta\lambda_3 - (\lambda_1 + \lambda_2) \leq 0.$$

By Lemma 9.7.2, λ_3 and $(-\lambda_1 - \lambda_2 - \lambda_3)$ are convex, and thus $[\delta\lambda_3 - (\lambda_1 + \lambda_2)]$ is convex.

We apply Theorem 9.6.1 in the situation described in Remark 9.7.1, with X the 0-sublevel set of the convex function $\Psi(e) = [\delta\lambda_3 - (\lambda_1 + \lambda_2)](e)$. The theorem tells us that the condition $[\delta\lambda_3 - (\lambda_1 + \lambda_2)] \leq 0$ is preserved under the Ricci flow (and hence that Ric $\geq \varepsilon Rg$ is preserved, as desired) provided that the same condition is preserved under the analogous ODE (9.6.1), or equivalently (9.7.1). Therefore, keeping in mind Remark 9.6.2, we must simply prove that

$$\frac{d}{dt}(\delta\lambda_3 - (\lambda_1 + \lambda_2)) \equiv 2\delta(\lambda_3^2 + \lambda_1\lambda_2) - 2(\lambda_1^2 + \lambda_2^2 + \lambda_2\lambda_3 + \lambda_1\lambda_3)$$

$$\leq 0,$$

(9.7.3)

whenever

$$\delta\lambda_3 - (\lambda_1 + \lambda_2) = 0. \tag{9.7.4}$$

In this regard, note that (9.7.4) implies that both $\lambda_3 \geq 0$ and $(\lambda_1 + \lambda_2) \geq 0$, since $\delta \in [0, 2)$ and $\lambda_1, \lambda_2 \leq \lambda_3$. Neglecting the easy case in which $\lambda_3 = 0$, (9.7.4) then implies that

$$\delta(\lambda_3^2 + \lambda_1\lambda_2) - (\lambda_1^2 + \lambda_2^2 + \lambda_2\lambda_3 + \lambda_1\lambda_3)$$

$$= \frac{1}{\lambda_3}\left[(\lambda_1 + \lambda_2)(\lambda_3^2 + \lambda_1\lambda_2) - \lambda_3(\lambda_1^2 + \lambda_2^2 + \lambda_2\lambda_3 + \lambda_1\lambda_3)\right]$$

$$= \frac{1}{\lambda_3}\left[\lambda_1^2\lambda_2 + \lambda_2^2\lambda_1 - \lambda_3(\lambda_1^2 + \lambda_2^2)\right]$$

$$\leq \frac{1}{\lambda_3}\left[\lambda_1^2\lambda_3 + \lambda_2^2\lambda_3 - \lambda_3(\lambda_1^2 + \lambda_2^2)\right]$$

$$= 0,$$

and hence (9.7.3) is satisfied. ☐

Corollary 9.7.6. *For all $\varepsilon \in [0, \frac{1}{3})$ and any Ricci flow $g(t)$ on a three-dimensional, closed manifold $\mathcal{M}$, the pinching condition*

$$\text{Ric} > \varepsilon Rg \tag{9.7.5}$$

is preserved.

Proof. By the condition (9.7.5) and the compactness of $\mathcal{M}$, we can find $\varepsilon' \in (\varepsilon, \frac{1}{3})$ such that Ric $\geq \varepsilon' Rg$. This condition is then preserved by Theorem 9.7.5. It remains to show that $\varepsilon' Rg > \varepsilon Rg$, or equivalently that $R > 0$, at each time t. This can be seen to be true at $t = 0$ by taking the trace of (9.7.5), and holds for subsequent times by Corollary 3.2.3. (We remark that Corollary 3.2.3 could also be derived from the ODE-PDE Theorem 9.6.1.) ☐

By setting $\varepsilon = 0$ in Theorem 9.7.5 and Corollary 9.7.6, we find the following special case.

Corollary 9.7.7.   *Under the Ricci flow on a three-dimensional closed manifold, the conditions* Ric ≥ 0 *and* Ric > 0 *are each preserved.*

Our next application of the ODE-PDE Theorem 9.6.1 is to see that the Ricci flow in three dimensions with Ric > 0 looks "round" where the scalar curvature is large.

Theorem 9.7.8 (Roundness).   *For any $0 < \beta < B < \infty$ and $\gamma > 0$ (however small) there exists $M = M(\beta, B, \gamma) < \infty$ such that whenever $g(t)$ is a Ricci flow on a closed, three-dimensional manifold, for $t \in [0, T]$, with $\beta g(0) \leq$ Ric$(g(0)) \leq Bg(0)$, we have*

$$\left| \text{Ric} - \frac{1}{3}Rg \right| \leq \gamma R + M,$$

for all $t \in [0, T]$.

Remark 9.7.9.   By estimating

$$\left| \text{Ric} - \frac{1}{3}Rg \right|^2 = |\text{Ric}|^2 - \frac{1}{3}R^2$$

$$= [(\lambda_2 + \lambda_3)^2 + (\lambda_1 + \lambda_3)^2 + (\lambda_1 + \lambda_2)^2] - \frac{1}{3}(2(\lambda_1 + \lambda_2 + \lambda_3))^2$$

$$= \frac{2}{3}(\lambda_1^2 + \lambda_2^2 + \lambda_3^2) - \frac{2}{3}(\lambda_1 \lambda_2 + \lambda_1 \lambda_3 + \lambda_2 \lambda_3)$$

$$= \frac{1}{3}[(\lambda_1 - \lambda_2)^2 + (\lambda_1 - \lambda_3)^2 + (\lambda_2 - \lambda_3)^2]$$

$$\leq (\lambda_3 - \lambda_1)^2,$$

we see that it suffices to control the weakly positive quantity $\lambda_3 - \lambda_1$.

Proof. (Theorem 9.7.8.)   At $t = 0$, we have Ric $\geq \beta g > 0$. In terms of the eigenvalues λ_i, this may be phrased

$$\lambda_1 + \lambda_2 \geq \beta > 0.$$

We define a first convex parallel function to be

$$\Psi_1 := \beta - (\lambda_1 + \lambda_2),$$

which will be weakly negative at $t = 0$. We also see readily that the weak negativity of Ψ_1 is preserved under the ODE (9.7.1).

Returning to $t = 0$, combining the facts $\beta g \leq$ Ric and $R \leq 3B$ tells us that Ric $\geq \frac{\beta}{3B}Rg$. After defining $\varepsilon = \frac{\beta}{3B} \in (0, \frac{1}{3})$, and $\delta = \frac{2\varepsilon}{1-2\varepsilon} \in (0, 2)$, we know from the proof of Theorem 9.7.5 that this condition is equivalent to

$$\Psi_2 := \delta\lambda_3 - (\lambda_1 + \lambda_2) \leq 0, \tag{9.7.6}$$

and is also preserved under the ODE (9.7.1).

Set $\theta = \frac{1}{1+\frac{1}{2}\delta} \in (\frac{1}{2}, 1)$. Note that at $t = 0$, $\lambda_3 - \lambda_1 \leq \lambda_3 - \lambda_1 + (\lambda_1 + \lambda_2) = \lambda_3 + \lambda_2 = \mathrm{Ric}(e_1, e_1) \leq B$ and $\lambda_1 + \lambda_2 = \mathrm{Ric}(e_3, e_3) \geq \beta$. Hence, by choosing $A > 0$ sufficiently large (depending on β and B) we have

$$\lambda_3 - \lambda_1 \leq A(\lambda_1 + \lambda_2)^\theta, \tag{9.7.7}$$

at $t = 0$.

We claim that (9.7.7) is preserved under our Ricci flow. If this were true, then by Remark 9.7.9 we would have

$$\left| \mathrm{Ric} - \frac{1}{3}Rg \right| \leq \lambda_3 - \lambda_1 \leq A(\lambda_1 + \lambda_2)^\theta \leq A(\lambda_1 + \lambda_2 + \lambda_3)^\theta = \tilde{A}R^\theta$$

for $\tilde{A} = \tilde{A}(\beta, B) := A\,2^{-\theta}$. The theorem would then hold with

$$M = M(\gamma, \beta, B) := \sup_{R \geq 0}(\tilde{A}R^\theta - \gamma R) < \infty.$$

It remains to show that (9.7.7) is preserved under this Ricci flow. By Lemma 9.7.2, $\lambda_3 - \lambda_1$ is convex, $\lambda_1 + \lambda_2$ is concave and $(\lambda_1 + \lambda_2)^\theta$ is concave where $\lambda_1 + \lambda_2 \geq 0$. Therefore,

$$\Psi_3 := (\lambda_3 - \lambda_1) - A(\lambda_1 + \lambda_2)^\theta$$

is a convex function where $\lambda_1 + \lambda_2 \geq 0$, and in particular, where $\Psi_1 \leq 0$.

We apply Theorem 9.6.1 in the situation described in Remark 9.7.1, with X consisting of all points where Ψ_1, Ψ_2 and Ψ_3 are all weakly negative. The theorem tells us that the set X is preserved under the Ricci flow (and in particular that (9.7.7) will continue to hold) provided that X is preserved under the analogous ODE (9.6.1), or equivalently (9.7.1).

We have already remarked that both the conditions $\Psi_1 \leq 0$ and $\Psi_2 \leq 0$ are preserved under the ODE (9.7.1). We will conclude by showing that if our eigenvalues solve (9.7.1) and satisfy $\Psi_1 \leq 0$ and $\Psi_2 \leq 0$, then whenever $\Psi_3 \geq 0$, the ratio $\frac{\lambda_3 - \lambda_1}{(\lambda_1 + \lambda_2)^\theta}$ is decreasing in time.

In this case, our eigenvalues satisfy $\lambda_3 - \lambda_1 > 0$ and $\lambda_1 + \lambda_2 > 0$, and we may calculate that

$$\frac{d}{dt}\ln(\lambda_3 - \lambda_1) = \frac{2}{\lambda_3 - \lambda_1}(\lambda_3^2 + \lambda_1\lambda_2 - \lambda_1^2 - \lambda_2\lambda_3) = 2(\lambda_3 - \lambda_2 + \lambda_1)$$

and that

$$\frac{d}{dt}\ln(\lambda_1 + \lambda_2) = \frac{2}{\lambda_1 + \lambda_2}(\lambda_1^2 + \lambda_2\lambda_3 + \lambda_2^2 + \lambda_1\lambda_3)$$

$$= 2\left[\lambda_3 + \frac{\lambda_1^2 + \lambda_2^2}{\lambda_1 + \lambda_2}\right] = 2\left[(\lambda_3 - \lambda_2 + \lambda_1) + \frac{2\lambda_2^2}{\lambda_1 + \lambda_2}\right].$$

Now $\delta\lambda_3 \leq \lambda_1 + \lambda_2$ (since $\Psi_2 \leq 0$) and $\lambda_1 \leq \lambda_2$, so

$$\delta(\lambda_3 - \lambda_2 + \lambda_1) \leq \lambda_1 + \lambda_2 = \frac{(\lambda_1 + \lambda_2)^2}{\lambda_1 + \lambda_2} \leq \frac{(2\lambda_2)^2}{\lambda_1 + \lambda_2}.$$

Hence

$$\frac{d}{dt}\ln(\lambda_1 + \lambda_2) \geq 2\left(1 + \frac{1}{2}\delta\right)(\lambda_3 - \lambda_2 + \lambda_1).$$

Thus

$$\frac{d}{dt}\ln\left[\frac{\lambda_3 - \lambda_1}{(\lambda_1 + \lambda_2)^\theta}\right] \leq 2(\lambda_3 - \lambda_2 + \lambda_1) - 2\theta\left(1 + \frac{1}{2}\delta\right)(\lambda_3 - \lambda_2 + \lambda_1)$$

$$= 2(\lambda_3 - \lambda_2 + \lambda_1)\left(1 - \theta\left(1 + \frac{1}{2}\delta\right)\right)$$

$$= 0,$$

by the definition of θ. Therefore, $\frac{\lambda_3 - \lambda_1}{(\lambda_1 + \lambda_2)^\theta}$ is decreasing as claimed, and (9.7.7) is preserved.   □

10
Three-manifolds with positive
Ricci curvature, and beyond

10.1 Hamilton's theorem

The estimates we saw in the previous chapter will help us understand the flow of a three-dimensional closed manifold with positive Ricci curvature. It turns out that such a manifold shrinks to nothing whilst gradually becoming round. Here we make this precise by taking a blow-up limit of the flow as we approach the first singular time (which we show must be finite) and arguing that it must converge to a round shrinking spherical space form which is diffeomorphic to the original manifold.

Theorem 10.1.1 (Hamilton). *If $(\mathcal{M}, g_0)$ is a closed three-dimensional Riemannian manifold with positive Ricci curvature, then the Ricci flow $g(t)$, with $g(0) = g_0$, on a maximal time interval $[0, T)$ becomes* round *in the following sense. There exist*

(i) a metric g_∞ on $\mathcal{M}$ of constant positive sectional curvature,
(ii) a sequence $t_i \uparrow T$,
(iii) a point $p_\infty \in \mathcal{M}$ and a sequence $\{p_i\} \in \mathcal{M}$,

such that if we define new Ricci flows $g_i(t)$ for $t \leq 0$ by

$$g_i(t) = |\mathrm{Rm}|(p_i, t_i)g\left(t_i + \frac{t}{|\mathrm{Rm}|(p_i, t_i)}\right) \qquad (10.1.1)$$

then

$$(\mathcal{M}, g_i(t), p_i) \to (\mathcal{M}, (c - t)g_\infty, p_\infty)$$

on the time interval $(-\infty, 0]$, for some $c > 0$.

Proof. First, let us pick $0 < \beta \leq B < \infty$ such that $\beta g_0 \leq \mathrm{Ric}(g_0) \leq Bg_0$. This is possible because $\mathcal{M}$ is compact and $\mathrm{Ric}(g_0) > 0$. Note then that $R \geq$

$3\beta > 0$ at $t = 0$, so by Corollary 3.2.4, we have $T \leq \frac{1}{2\beta}$, and in particular, there is a finite-time singularity in the flow.

We are then able to analyse the flow with Theorem 8.5.1, generating a constant $b > 0$, sequences $\{p_i\} \subset \mathcal{M}$ and $t_i \uparrow T$, rescaled Ricci flows $g_i(t)$ and (for $t \in (-\infty, b)$) a limit Ricci flow $(\mathcal{N}, \hat{g}(t))$ with a base point $p_\infty \in \mathcal{N}$.

Applying Theorem 9.7.8 to $g(t)$, we find that for all $t \in [0, T)$ and all $\gamma > 0$,

$$\left| \text{Ric} - \frac{1}{3} Rg \right| \leq \gamma R + M(\beta, B, \gamma),$$

with all terms computed with respect to $g(t)$. But $g_i(t)$ is just a rescaling, and translation in time, of $g(t)$, so with respect to $g_i(0)$ there holds

$$\left| \text{Ric} - \frac{1}{3} Rg \right| \leq \gamma R + \frac{M(\beta, B, \gamma)}{|\text{Rm}|(p_i, t_i)}$$

for all $\gamma > 0$. Taking the limit $i \to \infty$, we have

$$\left| \text{Ric} - \frac{1}{3} Rg \right| \leq \gamma R,$$

with respect to $\hat{g}(0)$. Since $\gamma > 0$ is arbitrary, we must have $\text{Ric} - \frac{1}{3} Rg = 0$ for $\hat{g}(0)$. Thus $\hat{g}(0)$ is Einstein, and R must be constant, as we discussed in Section 2.2.

Using the fact that $\mathcal{M}$ is three dimensional, $\hat{g}(0)$ then has constant sectional curvature (because the Einstein tensor $E = -\text{Ric} + \frac{1}{2} Rg = \lambda g$ for some $\lambda \in \mathbb{R}$). Moreover, since $R > 0$ for $g(t)$, it follows that $R > 0$ for $g_i(t)$ and so $R \geq 0$ for the limit $\hat{g}(0)$, and we deduce that the constant value of the sectional curvatures of $\hat{g}(0)$ is weakly positive. Recalling that $|\text{Rm}(\hat{g}(0))|(p_\infty) = 1$, we deduce that in fact $\hat{g}(0)$ must have strictly positive, constant sectional curvature.

By Myer's Theorem (which tells us that a positive lower bound for the Ricci curvature of a manifold implies that its diameter is finite – see [27, Theorem 11.8]) $\mathcal{N}$ must be compact and hence closed. By the definition of convergence of flows, we must then have $\mathcal{N} = \mathcal{M}$. By the backwards uniqueness of Ricci flows asserted in Theorem 5.2.2, we must have that $\hat{g}(t) = (c - t)g_\infty$ where g_∞ is some positive multiple of $\hat{g}(0)$. Therefore

$$(\mathcal{M}, g_i(t), p_i) \to (\mathcal{M}, (c - t)g_\infty, p_\infty).$$

$\square$

Merely the existence of a metric such as g_∞ gives an important corollary.

Corollary 10.1.2. *Any closed Riemannian three-manifold with positive Ricci curvature admits a metric of constant positive sectional curvature. In particular, if the manifold is also simply connected, then it must be S^3.*

10.2 Beyond the case of positive Ricci curvature

From here, we would like to drop the Ric > 0 hypothesis and prove Thurston's Geometrisation Conjecture by understanding singularities and performing surgeries as discussed in Section 1.5. It is important to develop an understanding of general blow-ups of singularities, and the first key tool in this direction is the following curvature pinching result of Hamilton and Ivey ([25], [21]) which is readily proved using Theorem 9.6.1.

Theorem 10.2.1 (Hamilton-Ivey). *For all $B < \infty$ and $\varepsilon > 0$, there exists $M = M(\varepsilon, B) < \infty$ such that if $(\mathcal{M}, g_0)$ is a closed three-manifold with $|\mathrm{Rm}(g_0)| \leq B$, then the Ricci flow $g(t)$ with $g(0) = g_0$ satisfies*

$$E + (\varepsilon R + M)g \geq 0, \qquad (10.2.1)$$

where E represents the Einstein tensor, as before.

This means that we can estimate the lowest sectional curvature by

$$\lambda_1 \geq -\varepsilon R - M,$$

and since $R = 2(\lambda_1 + \lambda_2 + \lambda_3)$, this means that if the flow has a very negative sectional curvature, then the most positive sectional curvature is much larger still.

Corollary 10.2.2. *For a Ricci flow on a closed three-dimensional manifold, when we take a blow-up limit*

$$(\mathcal{M}, g_i(t), p_i) \to (\mathcal{N}, \hat{g}(t), p_\infty),$$

as in Theorem 8.5.1, the manifold $(\mathcal{N}, \hat{g}(t))$ has weakly positive sectional curvature for all $t \in (-\infty, b)$.

Remark 10.2.3. This severely constraints the singularities which are possible.

To understand more about the Ricci flow from where these notes end, the reader is directed first to the paper of Perelman [31].

Appendix A

Connected sum

In this appendix, we recall the notion of connected sum, as required principally in Section 1.4.

Given any two oriented manifolds A and B of the same dimension, we first remove a small disc/ball in each.

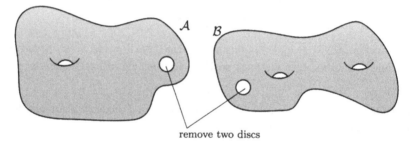

remove two discs

We then paste them together (matching orientations) to get a new manifold which we denote by $A\#B$.

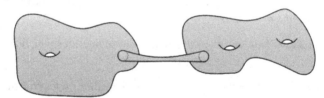

For example, $T^2\#T^2$ is shown in figure A.1.

genus $= 2$

Figure A.1 $T^2\#T^2$.

References

[1] S. ANGENENT and D. KNOPF, *An example of neckpinching for Ricci flow on* S^{n+1}. Math. Res. Lett. **11** (2004) 493–518.

[2] A. L. BESSE, Einstein manifolds. 'A series of modern surveys in mathematics.' Springer-Verlag, 1987.

[3] D. BURAGO, Y. BURAGO and S. IVANOV, A course in metric geometry. 'Graduate Studies in Math.' **33** A.M.S. 2001.

[4] I. CHAVEL, Riemannian geometry: a modern introduction. 'Cambridge tracts in math.' **108** C.U.P. 1995.

[5] B.-L. CHEN and X.-P. ZHU, *Uniqueness of the Ricci flow on complete noncompact manifolds.* http://arXiv.org/math.DG/0505447v3 (2005).

[6] X.-X. CHEN, P. LU and G. TIAN, *A note on uniformization of Riemann surfaces by Ricci flow* http://arXiv.org/math.DG/0505163v2 (2005).

[7] B. CHOW and D. KNOPF, The Ricci flow: An introduction. 'Mathematical Surveys and Monographs.' **110** A.M.S. 2004.

[8] T. H. COLDING and W. P. MINICOZZI, *Estimates for the extinction time for the Ricci flow on certain 3-manifolds and a question of Perelman.* http://arXiv.org/math.AP/0308090v2 (2003).

[9] D. DETURCK, *Deforming metrics in the direction of their Ricci tensors.* In 'Collected papers on Ricci flow.' Edited by H. D. Cao, B. Chow, S. C. Chu and S. T. Yau. Series in Geometry and Topology, 37. International Press, 2003.

[10] J. EELLS and L. LEMAIRE, *A report on harmonic maps.* Bull. London Math. Soc. **10** (1978) 1–68.

[11] J. EELLS and J. H. SAMPSON, *Harmonic mappings of Riemannian manifolds.* Amer. J. Math. **86** (1964) 109–169.

[12] K. FUKAYA, *A boundary of the set of the Riemannian manifolds with bounded curvatures and diameters.* J. Differential Geom. **28** (1988) 1–21.

[13] D. GABAI, A.I.M. notes. http://www.aimath.org/WWN/geometrization

[14] S. GALLOT, D. HULIN and J. LAFONTAINE, Riemannian geometry. (Second edition) Springer-Verlag, 1993.

[15] D. GILBARG and N. S. TRUDINGER, Elliptic Partial Differential Equations of Second Order. (Second edition.) Springer-Verlag, 1983.

110 *References*

sgeom. 17 (1982) 255–306.

[20]   R. S. HAMILTON, *Four-manifolds with positive curvature operator.* J. Differential geom. 24 (1986) 153–179.

[21]   R. S. HAMILTON, *The formation of singularities in the Ricci flow.* Surveys in differential geometry, Vol. II (Cambridge, MA, 1993) 7–136, Internat. Press, Cambridge, MA, 1995.

[22]   R. S. HAMILTON, *A compactness property for solutions of the Ricci flow.* Amer. J. Math. 117 (1995) 545–572.

[23]   R. S. HAMILTON, *Four-manifolds with positive isotropic curvature.* Comm. Anal. Geom. 5 (1997) 1–92.

[24]   A. HATCHER, Notes on basic 3-manifold topology. http://www.math. cornell.edu/~hatcher

[25]   T. IVEY, *Ricci solitons on compact three-manifolds.* Diff. Geom. Appl. 3 (1993) 301–307.

[26]   M. KAPOVICH, 'Thurston's geometrization and its consequences.' MSRI video, (2003). http://www.msri.org/publications/ln/msri/ 2003/ricciflow/kapovich/1/index.html

[27]   J. M. LEE, Riemannian manifolds. 'Graduate texts in math.' 176 Springer-Verlag, 1997.

[28]   P. LI and S.-T. YAU, *On the parabolic kernel of the Schrödinger operator.* Acta Math. 156 (1986) 154–201.

[29]   J. MILNOR, Morse theory. 'Annals of Mathematical Studies,' 51. Princeton University Press, 1969.

[30]   L. NI, *The entropy formula for linear heat equation.* J. Geom. Anal. 14 (2004) 87–100; *Addendum.* J. Geom. Anal. 14 (2004) 369–374.

[31]   G. PERELMAN *The entropy formula for the Ricci flow and its geometric applications.* http://arXiv.org/math.DG/0211159v1 (2002).

[32]   G. PERELMAN *Ricci flow with surgery on three-manifolds.* http://arXiv. org/math.DG/0303109v1 (2003).

[33]   G. PERELMAN *Finite extinction time for the solutions to the Ricci flow on certain three-manifolds.* http://arXiv.org/math.DG/0307245v1 (2003).

[34]   O. S. ROTHAUS, *Logarithmic Sobolev inequalities and the spectrum of Schrödinger operators.* J. Funct. Anal. 42 (1981) 110–120.

[35]   P. SCOTT, *The geometries of three-manifolds.* Bull. L.M.S. 15 (1983) 401–487.

[36]   W.-X. SHI, *Deforming the metric on complete Riemannian manifolds.* J. Differential Geom. 30 (1989) 223–301.

[37]   M. SIMON, *A class of Riemannian manifolds that pinch when evolved by Ricci flow.* Manuscripta math. 101 (2000) 89–114.

[38]   W. P. THURSTON, Three-dimensional geometry and topology, volume 1. Princeton University Press, 1997.

[39] P. M. TOPPING, *Diameter control under Ricci flow* Comm. Anal. Geom. **13** (2005).

[40] P. M. TOPPING, *Ricci flow compactness via pseudolocality, and flows with incomplete initial metrics.* Preprint (2006).

[41] S. R. S. VARADHAN, Probability Theory. 'Courant Lect. Notes in Math.' **7** A.M.S. 2001.

[42] C. VILLANI, Topics in optimal transportation. 'Graduate Studies in Math.', **58** A.M.S. 2003.

[43] R. YE, *Global existence and uniqueness of Yamabe flow.* J. Differential geom. **39** (1994) 35–50.

Index